The Anthropocene, Ontopolitics, and International Relations

Matthew Robson
Editor

The Anthropocene, Ontopolitics, and International Relations

Springer

Editor
Matthew Robson
Faculty of Political Science and Public
Administration
Chiang Mai University
Chiang Mai, Thailand

ISBN 978-981-96-2924-4 ISBN 978-981-96-2925-1 (eBook)
https://doi.org/10.1007/978-981-96-2925-1

This Springer imprint is published by the registered company Springer Nature Singapore Pte Ltd.
The registered company address is: 152 Beach Road, #21-01/04 Gateway East, Singapore 189721, Singapore

Foreword

The Anthropocene offers a challenge to mainstream approaches to International Relations and to thinking across a range of policy concerns. This challenge is posed in several distinct registers, well expressed in this diverse collection, which emphasizes the ontopolitical dimensions at stake. Ontopolitics, as the introductory chapter states, brings to the forefront the need to interrogate 'the often-unrecognized ontological assumptions which underpin theories, beliefs and practices' of agents and actors seeking to understand the Anthropocene as a problematic and to develop new approaches to governance and policymaking today. Exemplary here are conceptions of the 'posthuman', of multi-species and new and entangled forms of security, and of the 'Capitalocene', 'Chthulucene', 'Plantationocene', 'Technocene', and 'Kinocene'—just to highlight some of the differing concerns and conceptual lenses well-represented in this collection. Each of the chapters deploys a different framing and the ontopolitical assumptions and ethical and political stakes are drawn out in the chapters themselves and in the editor's final 'Conclusions' chapter.

This edited collection brings to the fore the fact that the most important aspect of the Anthropocene for scholars of International Relations, familiar already with issues of environmental security or with discourses of interdependency and interrelation, is the ontological register of much of the discussion. The Anthropocene is ontological in the sense that it is not so much a new problem to be confronted as it is the recognition that something has already fundamentally ended. That something is the world of modernity in which humans were imagined as rational, autonomous subjects and 'nature' as somehow separate, suborned to universal laws of causality. Modernity itself, we are now told, was never how we understood it to be and, as Dipesh Chakrabarty argues, its futural promises are now non-fulfillable.[1] The sense of limits is palpable in these discourses of closure but, as this collection emphasizes, alternative worlds are also possible, and the ethical and political stakes could not be higher.

Bruno Latour has pointed out that modernity was a paradoxical condition: The more that we imagined ourselves as subjects separated from the world, developing

[1] D. Chakrabarty (2021) *The Climate of History in a Planetary Age*, University of Chicago Press.

knowledge of how we could direct and control 'natural' processes, the more humanity grew entangled within these processes. Modernity itself was the midwife to processes that were no longer 'natural' nor amenable to external control or direction by human subjects, in spite (or because) of the fact that the latter were seen to have all the powers of agency while the rest of the world—of nonhumans—was seen to be merely passive objects of our intentionality.[2] As Timothy Morton argues, the awareness of human-induced climate change and of our dependence upon nonhuman agency has done what continental philosophy and the radical ideas of postmodernism failed to do, i.e., to fundamentally question our modernist knowledge assumptions.[3]

It is important to appreciate that much of the Anthropocene discourse challenges the epistemological and ontological framings of modernity from a position that is pro—rather than anti-science. For Morton, the more we know about radiation, global warming, and other intractable problems that we face, the more their anthropogenic nature becomes apparent; thus, we increasingly appear to be part of the problem rather than the solution. He terms these objects, which cannot be separated from us, as observers, 'hyperobjects'.[4] William Connolly focuses on how the geo sciences reveal that the Earth's 'planetary force fields'—such as climate patterns, ocean conveyor systems, species evolution, glacier flows, and air circulations—exhibit self-organizing capacities that can go through volatile and rapid changes.[5] Thus, the Anthropocene is not new, except in the fact that human impacts amplify the non-linear and interactive effects of these forces in increasingly unpredictable ways.

Perhaps the most important aspect of this ontological engagement with the Anthropocene is that often work is not so much about saving the world from the climate change emergency but how we practice different ways of being in the world after the end of the modernist imaginary of progress. Exemplary here is Anna Lowenhaupt Tsing's work, *The Mushroom at the End of the World: On the Possibility of Life in Capitalist Ruins*.[6] Her starting assumption is the end of the modern division between humanity and nature. She argues that without this artificial separation, all creatures can be given care and respect, and we are free to express ourselves without the constraints of a 'parochial' rationality. The Anthropocene thus enables us to think 'after failure', 'after progress', 'after the end of the world'.

In this way, the Anthropocene can be understood as a world that is fuller, livelier, and more entangled than the soulless, simplified, and atomized world of modernity. In the ruins of modernity there is more life than could possibly have been imagined by modernist human subjects convinced of their separation from the world. Realizing our precarious condition brings us back to the world: the Anthropocene is like an unseen

[2] B. Latour (1993) *We Have Never Been Modern*, Harvard University Press.

[3] T. Morton (2013) *Hyperobjects: Philosophy and Ecology after the End of the World*, University of Minnesota Press.

[4] Ibid.

[5] W. E. Connolly (2017) *Facing the Planetary: Entangled Humanism and the Politics of Swarming*, Duke University Press.

[6] A. L. Tsing (2015) *The Mushroom at the End of the World: On the Possibility of Life in Capitalist Ruins*, Princeton University Press.

force, imposing a new sociability and new set of sensitivities on the basis that we are no longer separate, no longer in control, no longer not interested in other actors and agencies with which we cohabit. The Anthropocene is thereby not merely a world of doom and gloom and extinction but also an invitation to be curious, imaginative, and exploratory, developing new forms of technological thought, new modes of political struggle and accountability, new approaches to human and more-than-human justice and governance and to ethical, socio-economic and political practices of community.

David Chandler
University of Westminster
London, UK

Contents

Contributors

Ana Flávia Barros-Platiau University of Brasilia (UnB), Brasília, Brazil

Fábio Henrique Granja e Barros Federal Accounting Court (TCU), Brasilia, Brazil

Paulo Eduardo Saraiva Câmara University of Brasilia (UnB), Brasília, Brazil; Brazilian Defence College (ESD), Brasilia, Brazil

Guilherme Lopes da Cunha Brazilian War College (ESG), Rio de Janeiro, Brazil

Inês Ferreira de Sousa NOVA University of Lisbon, Lisbon, Portugal

Jorge Gomes do Cravo Barros Brasilia, Brazil

Gabriela Kuetting Rutgers University, Newark, NJ, USA

Yoan Molinero-Gerbeau Instituto Universitario de Estudios sobre Migraciones, Universidad Pontificia Comillas, Madrid, Spain

Joana Castro Pereira University of Porto, Porto, Portugal

Victor Pérez-Segura Instituto Universitario de Estudios sobre Migraciones, Universidad Pontificia Comillas, Madrid, Spain

Matthew Robson Faculty of Political Science and Public Administration, Chiang Mai University, Chiang Mai, Thailand

Álvaro San Román Gómez Universidad Nacional de Educación a Distancia, Madrid, Spain

Carlos Henrique Rubens Tomé Silva University of Brasilia (UnB), Brasília, Brazil

João Terrenas University of Beira Interior, Covilhã, Portugal

Laura Cristina Feindt Urrejola University of Brasilia (UnB), Brasília, Brazil; Universidad Nacional de La Plata, Buenos Aires, Argentina

Chapter 1
Introduction: The Anthropocene, Ontopolitics, and International Relations

Matthew Robson

International Environmental Policy in the 'Anthropocene'

The 2023 'Synthesis Report' of the Intergovernmental Panel on Climate Change (IPCC) was as disheartening as it was predictable. Among a plethora of damning statements, it affirms that 'human activities, principally through emissions of greenhouse gases, have unequivocally caused global warming,' and that 'widespread and rapid changes in the atmosphere, ocean, cryosphere and biosphere have occurred.' The report continues, 'human-caused climate change is already affecting many weather and climate extremes in every region across the globe,' something which 'has led to widespread adverse impacts and related losses and damages to nature and people.'[1] According to projections, the future is, seemingly, no more promising given that breaching the oft-cited 1.5 °C threshold for global temperature rises is becoming increasingly likely: 'Global GHG emissions in 2030 implied by nationally determined contributions (NDC's) announced in October 2021 make it likely that warming will exceed 1.5 °C during the twenty-first century and make it harder to limit warming below 2 °C.'[2] The devastating impact and irreversible consequences of these processes on the Earth's diverse life forms further amplify anxieties about the direction we are heading. That such reports may even be conservative in their assessments is another disquieting consideration.[3]

---

[1] IPCC (2023) 'Summary for Policymakers,' in *Climate Change 2023: Synthesis Report*, IPCC: Geneva, pp. 1–34.

[2] Ibid.

[3] S. Herrando-Pérez et al. (2019) 'Statistical Language Backs Conservatism in Climate-Change Assessments,' *BioScience*, 69(3), pp. 209–219.

M. Robson (✉)
Faculty of Political Science and Public Administration, Chiang Mai University, Chiang Mai, Thailand
e-mail: matthew.r@cmu.ac.th

M. Robson (ed.), *The Anthropocene, Ontopolitics, and International Relations*,
https://doi.org/10.1007/978-981-96-2925-1_1

Within this context, international institutions have proposed a combination of both mitigation and adaptation to tackle the growing environmental challenges and address the ensuing ecological disaster. For the most part, these measures have revolved around sustainable development-resilience conceptual frameworks. While notions of sustainable development have been around for some decades,[4] the concept has risen incrementally to the forefront of the policy agendas of international institutions like the United Nations and the World Bank. The United Nations Development Programme (UNDP) has been aiming to advance a broad range of traditional development goals within an environmentally conscious schema oriented toward a 'better and more sustainable future for all.'[5] Building 'resilience,' itself now ubiquitous in manifold policy sectors, has been increasingly understood as a key precursor for sustainable development programs amid changing climatic and environmental conditions.[6] With its roots in the field of Ecology, 'resilience' has assumed a prominent position in international development and environment policy discourse, where it broadly denotes the capacities of individuals, groups, and communities to adapt to, and thus withstand, external challenges and shocks.[7]

International environmental policy frameworks have converged in recent times with academic discourse around the acknowledgment of the 'Anthropocene' as the key signifier designating our contemporary historical period.[8] This term, elaborated from within the natural sciences, denotes the transition to a new geological epoch ushered in from the impact that human beings have had on the Earth System.[9] Thus, as Clive Hamilton insists, it is human disruption of the planet Earth as an interconnected singularity or totality which the 'Anthropocene' specifically names.[10] Major international institutions of global governance have recently incorporated the concept into their policy frameworks as the overriding context within which human socio-political and economic activity is now taking place. In its annual Human Development Report, for instance, the United Nations Development Program (UNDP) titled their publication of 2020 'The Next Frontier—Human Development and the Anthropocene.'[11] The report clearly considers the 'Anthropocene' to be an unprecedented phase in the

---

[4] G. H. Brundtland. & M. Khalid (1987) *World Commission on Environment and Development: Our Common Future*, New York: Oxford University Press.

[5] United Nations (2024) *Sustainable Development Goals*, UN.org, available at: https://www.un.org/sustainabledevelopment/sustainable-development-goals/.

[6] T. Chaigneau et al. (2022) 'Reconciling well-being and resilience for sustainable development,' *Nature Sustainability*, 5, pp. 287–293.

[7] J. Walker & M. Cooper (2011) 'Genealogies of resilience: From systems ecology to the political economy of crisis adaptation,' *Security Dialogue*, 42, pp. 155–156.

[8] In this introductory chapter the 'Anthropocene' shall be left in quotation marks to emphasize the contested nature of the term.

[9] P. J. Crutzen (2002) 'Geology of Mankind,' *Nature*, 415, 3rd January 2002, p. 23.

[10] C. Hamilton (2020) 'Towards a fifth ontology for the Anthropocene,' *Angelaki: Journal of the Theoretical Humanities*, 25(4), pp. 110–119.

[11] United Nations Development Program (2020) *The next frontier—Human development and the Anthropocene*, Human Development Report.

history of humankind: 'For the first time in our history the most serious and immediate, even existential, risks are human made and unfolding at planetary scale.'[12] In a later report published in 2022, the UNDP sets out to map the 'New threats to human security in the Anthropocene.'[13] Elsewhere, the concept 'Anthropocene' has also featured in other policy reports and briefings such as those of the World Bank and the European Union.[14]

At the same time, the emergence of the 'Anthropocene' poses a significant challenge to existing environmental policy frameworks. For one thing, the implied collapse of the culture-nature binary suggests that, strictly speaking, the very notion of an 'environmental' policy is misconceived.[15] As Frank Biermann notes, this 'paradigm' no longer seems appropriate, since, among other things, it 'emphasizes a dichotomy of 'humans' and 'nature' that is no longer defensible.'[16] The 'Anthropocene' also calls into question the hegemonic rationalities and technologies of governance which populate these policy frameworks. While authors within International Relations (IR) had already been expressing their dissatisfaction with the pursuit of sustainable development and resilience,[17] the implications of the 'Anthropocene' seem to bring some of these issues to the fore. Authors have directed at least two principal charges against 'resilience' as a way of thinking about governance in the 'Anthropocene.' First, although resilience approaches presume that life is 'in excess of being,' it does not necessarily follow that this power can be harnessed by human beings: 'the fact that life withholds excessive powers of actualization does not immediately equate to these powers as being generative and productive 'for us,' nor to the possibility that humans can have the capacity to understand, instrumentalize or to actualize these powers.'[18] Indeed, acknowledgment that the distribution of force between human beings and nature in the 'Anthropocene' is 'asymmetric,' with the latter possessing forces which surpass immeasurably the capacities of the former, building resilience would amount to little more than a futile exercise in 'fire-fighting'

[12] Ibid., p. 21.

[13] United Nations Development Program (2022) 'New threats to human security in the Anthropocene: Demanding greater solidarity,' *Special Report*, UNDP.

[14] World Bank (2022) 'Change Cannot Wait: Building Resilient Health Systems in the Shadow of COVID-19 -Investing in Health System Resilience for the Anthropocene,' Washington D.C.: World Bank Group; European Environment Agency (2023) 'Exiting the Anthropocene? Exploring fundamental change in our relationship with nature,' Briefing no. 24.

[15] F. Biermann (2022) 'The future of 'environmental' policy in the Anthropocene: time for a paradigm shift' in G. Hayes et al. (Eds) *Trajectories in Environmental Politics*, London: Routledge.

[16] Ibid., p. 58.

[17] See for instance J. Joseph (2013) 'Resilience as embedded neoliberalism: a governmentality approach,' *Resilience: International Policies, Practices and Discourses*, 1(1), p. 42; J. Reid (2014) 'Interrogating the Neoliberal Biopolitics of the Sustainable Development-Resilience Nexus,' *International Political Sociology*, 7(4), pp. 353–367; B. Evans & J. Reid (2014) *Resilient Life*, Polity; D. Chandler (2014) *Resilience: The Governance of Complexity*, Abingdon: Routledge; J. Reid (2016) 'Resilience: The biopolitics of security,' in D. Chandler & J. Reid (Eds.) *The neoliberal subject: Resilience, adaptation and vulnerability*, London: Rowman & Littlefield, pp. 51–72.

[18] S. Wakefield, D. Chandler, and K. Grove (2021) 'The asymmetrical anthropocene: resilience and the limits of posthumanism,' *Cultural Geographies*, 29(3), p. 7.

(if the principal drivers of the climate crisis are not dealt with in any meaningful way).[19] A second and related issue is the way in which resilience thinking typically privileges 'relations over entities.'[20] Since the relations which make up a particular society or community are deemed to be irrevocably exposed to threats emanating from the natural world (these are deemed to be effects of the relations themselves), it thus becomes a question of merely responding to the effects, which, importantly, can in turn only be done through adaptation. In short, there is 'the assumption that shocks are 'inevitable' and that adaptation is the only way forward.'[21] The 'Anthropocene,' with its emphasis on the agential capacities of human beings in altering the Earth System, exposes the folly of this way of thinking by re-routing us back to a concern with causes and not merely effects. It turns out that resilience thinking is profoundly depoliticizing and ultimately ineffective since it seeks 'to continue to preserve and reproduce exactly the same modes of being and thinking that caused anthropogenic climate change in the first place.'[22]

The 'Anthropocene' and International Relations (IR)

To grasp the fundamental challenge that the 'Anthropocene' poses to the disciplinary foundations of IR some authors have preferred to think of it as not so much a 'problem,' but rather a 'condition that we are in.'[23] As the concept brings into focus not only the constitutive role of human beings in altering the Earth System, but more importantly the immense agential capacities of the latter in affecting the former, the ontological and epistemological architecture of IR appears to be crumbling around it. This is a discipline, like others in the social sciences, which has traditionally been structured around binary oppositions demarcating a 'subject' from an 'object', an 'inside' from an 'outside,' a 'political/social world' from a 'natural world,' and so forth. It is indeed through these presuppositions that essentialist and foundationalist conceptions of the 'state,' 'international system,' and 'security' have been formulated. Yet such thinking, which for some descends from a Cartesian philosophical tradition now in its death throes, is rendered problematic by a set of natural processes which are critically impacting upon humankind's social-economic-political existence. Faced with transforming climatic and environmental conditions, the complexities of a radically interconnected world seem more apparent, as the natural world refuses its relegated status as passive backdrop to the troubles and tribulations played

[19] D. Chandler (2020) 'Security through societal resilience: Contemporary challenges in the Anthropocene,' *Contemporary Security Policy*, 41(2), p. 195.

[20] D. Chandler, K. Grove and S. Wakefield (Eds.) (2020) *Resilience in the Anthropocene: Governance and Politics at the End of the World*, London: Routledge, pp. 12–15,

[21] Ibid., p. 14.

[22] Ibid., p. 14.

[23] D. Chandler, F. Muller, & D. Rothe (Eds.) (2021) *International Relations in the Anthropocene: New Agendas, New Agencies and New Approaches*, Palgrave Macmillan, p. 2.

out by different socio-political formations. For some scholars of IR, therefore, what is required is no less than a re-appraisal and radical re-configuration of the ontological, epistemological, and ethical foundations of the discipline.[24]

The 'Anthropocene' rupture has indeed shaken up several research areas which are considered staples in the field of IR.[25] First, the implications of this seemingly new geological age have led to a range of critical studies on governance at the international and global scales.[26] In a notable contribution, Burke et al. (2016) have made a call for novel forms of governance as part of a 'Planet Politics,' which they conceive of as coming from the 'End of IR.'[27] Not everyone in the field has been receptive to the 'manifesto' presented though, with some authors reproaching it for what they deem to be a return to a now discredited liberal cosmopolitanism and 'top-down' global governance model.[28] What is required, conversely, is less top-down governance and more bottom-up resistance.

The impact of the concept 'Anthropocene' on IR has also extended to the field of security. Mainstream IR theories positing *national* security as the primary concern of states—often conceived in conjunction with an anarchical international system—have become increasingly problematic in a world where the threats posed by the planet itself transcend those pertaining to other state entities. Perhaps more than ever, it has become crucial to interrogate the different discourses of 'climate security' and the actions they make possible.[29] When we do just that, even purportedly 'critical' discourses aimed at rethinking security in this shifting environmental context can be called into question. Madeleine Fagan contends that calls for 'environmental security' are fundamentally flawed due to their reliance on conceptions of the 'environment' and 'security' that perpetuate the outdated binaries detaching humans from non-humans, and culture from nature.[30] 'Ecological security' suffers a similar fate, argues Fagan, as the shift to ecology merely 'reinscribes and obscures' the distinction between culture and nature. Neither environmental nor ecological security, therefore, seem to escape from the ontological and epistemological strictures which have shaped classical accounts of security in IR. The significance of all this hinges on the limits

[24] C. Harrington (2016) 'The Ends of the World: International Relations and the Anthropocene,' *Millennium: Journal of International Studies*, 44(3), pp. 478–498; A. Burke et al. (2016) 'Planet Politics: A Manifesto from the End of IR,' *Millennium: Journal of International Studies*, 44(3), pp. 1–25.

[25] Chandler, Muller & Rothe 'International Relations in the Anthropocene.'

[26] See for instance A. Flávia Barros-Platiau, E. Viola, & M. Franchini (2017) 'The Challenges of the Anthropocene: From International Environmental Politics to Global Governance,' *Scielo: Ambiente & Sociedade*, 20(3), pp. 177–199.

[27] Burke et al., 'Planet Politics'; See also, F. Biermann (2014) 'The Anthropocene: A Governance Perspective,' *The Anthropocene Review*, 1(1), pp. 57–61.

[28] D. Chandler, E. Cudworth, and S. Hobden (2017) 'Anthropocene, Capitalocene and Liberal Cosmopolitan IR: A Response to Burke et al.'s 'Planet Politics,' *Millennium: Journal of International Studies*, 46(2), pp. 190–208.

[29] M. McDonald (2013) 'Discourses of Climate Security,' *Political Geography*, 33, pp. 42–51.

[30] M. Fagan (2017) 'Security in the Anthropocene: Environment, ecology, escape,' *European Journal of International Relations*, 23(2), pp. 292–314.

which are then placed on the 'potential of the concept of 'Anthropocene' to offer a critical framework with which to analyse the interplay of nature and culture in contemporary security politics.'[31] Meanwhile, as Cameron Harrington has pointed out, there is a discernable shift in focus toward 'a growing alertness to the diverse forms of life that produce and are affected by conditions of (in)security.'[32] In short, for some authors any dualisms separating a sovereign and rational human world from a passive and malleable natural world are no longer deemed appropriate to account for the security threats posed to humans, non-humans, and inanimate bodies alike.

The 'Anthropocene' has also been a subject of discussion and exploration within the sub-field of geopolitics. While authors have shown an interest in the links between geopolitics and the environment for quite some time, the emergence of the concept 'Anthropocene' has marked an important point of inflection in these studies.[33] The 'classical' world of geopolitics that pits states in competition and conflict with one another over space and natural resources, one which is predicated on a spacialization of world politics composed of quintessentially stable and discrete territorial state entities, is difficult to reconcile with the dissolution of the culture-nature binary suggested by the notion 'Anthropocene.'[34] Of course, neither carbon emissions nor the planetary phenomenon of global heating they give rise to have ever respected the artificial territorial boundaries still reified in geopolitical commentary today.[35] One can only wonder as geopolitical competition over a melting Arctic gathers pace, and we stand at the brink of an artificially produced ecological catastrophe.[36] The time for novel forms of geopolitics is long overdue, one which can emerge from the continuous critique of established geopolitical categories, and which cultivates the innate human capacity for cooperation as opposed to conflict.[37]

The 'Anthropocene' has also disrupted migration research within IR. Discussions on the environment and migration are often prone to environmental determinism, associating environmental decline with visions of migrant 'floods' flowing across borders. While it is to be expected that vulnerable people would seek to find safer and more habitable ground—with the so-called 'developed world' being the most likely destination—reducing the phenomenon of 'climate migration' to environmental factors risks effacing the constitutive roles of human beings in bringing

[31] Ibid., p. 292.

[32] C. Harrington (2017) 'Posthuman Security and Care in the Anthropocene,' in *Reflections on the Posthuman in International Relations: The Anthropocene, Security and Ecology* Eds. C. Eroukhmanoff & M. Harker, E-International Relations, p. 73.

[33] S. Dalby (2007) 'Anthropocene Geopolitics: Globalisation, Empire, Environment and Critique,' *Geography Compass*, 1(1), pp. 103–118.

[34] Ibid, p. 111.

[35] Ibid, p. 4.

[36] M. Gross (2020) 'Geopolitical Competition in the Arctic Circle,' *Harvard International Review*, 2nd December, 2020.

[37] Dalby, 'Anthropocene Geopolitics,' pp. 115–116.

about the crisis in the first place.[38] In so far as human responsibility logically evaporates along with human agency, there are important ethical considerations in play here as well. One can envisage how the mobilization of environmentally deterministic arguments might serve the objectives of certain political groups with an interest in absolving themselves of responsibility for stricken Others. Another intriguing line of research emerges when we consider 'climate migration' as a field of political struggle, and not merely through the lens of liberal humanitarianism or a politics of 'the human.'[39] Vicky Squire has argued that it is the governing practices of migration themselves, embedded within 'racialized dynamics of colonial violence,' which *produce* migration as precarious and a 'global challenge' to be solved in the first place.[40] There are in fact several shortcomings with the concept 'climate migration' (or 'climate refugee'), meaning that many authors have instead been turning to the category of 'mobility.'[41] Moreover, this turn to mobility is influenced by the concept 'Anthropocene' in so far as it 'draws us into ever closer proximity to Earth's geomorphology, its dynamism, its fluidity, the inherent mobility of the Earth system.'[42]

The notion of an 'Anthropocene,' and that of anthropogenic climate change, has prompted authors within IR and beyond to explore its ethical implications. Ethics has indeed been a key concern in international environmental and climate politics for some time as a means of guiding political decision-making and exploring more just ways of dealing with the crisis.[43] While liberal ethico-political approaches have perhaps been most prominent in this body of work, with a particular emphasis on liberal conceptions of 'justice,'[44] some have questioned whether or not this translates into policy prescription.[45] Others have charged these ethico-political frameworks with typically failing to diagnose and address the unjust and noxious systemic conditions of global capitalism which are driving climate change.[46] Yet the emergence of the concept 'Anthropocene,' along with its problematization of the human-nature binary opposition, has also cast a shadow over such humanist ethical frameworks.

[38] E. Lunstrum & P. S. Bose (2022) 'Environmental displacement in the Anthropocene,' *Annals of the American Association of Geographers*, 112(3), pp. 644–653.

[39] V. Squire (2020) 'Migration and the politics of 'the human': confronting the privileged subjects of IR,' *International Relations*, 34(3), pp. 290–308.

[40] Ibid, p. 290.

[41] A. Baldwin, C. Frohlich, & D. Rothe (2019) 'From Climate Migration to Anthropocene Mobilities: Shifting the Debate,' *Mobilities*, 14(3), pp. 289–297.

[42] Ibid.

[43] See for instance S. Gardiner (2009) 'Ethics and Climate Change: An Introduction,' WIREs Climate Change, 1(1), 54–66.

[44] G. Calder & C. McKinnon (Eds.) (2012) *Climate Change and Liberal Priorities*, Routledge: London.

[45] R. Falkner (2019) 'The Unavoidability of Justice—and Order—in International Climate Politics: From Kyoto to Paris and Beyond,' *British Journal of Politics and International Relations*, 21.

[46] D. Faber, B. Levy & C. Schlegel (2021) 'Not all People are Polluted Equally in Capitalist Society: An Eco-Socialist Commentary on Liberal Environmental Justice Theory,' *Capitalism Nature Socialism*.

For instance, we could argue that liberal ethical approaches engender the reification of a now defunct self-contained and sovereign humanist subject. Such notions would appear to be critically undermined by an ecological context in which the agential capacities of the natural world are coming sharply into focus, demonstrating its refusal to adhere to its designated status as mere background to human affairs. Assuming 'posthuman' ontologies authors have been proposing alternative ethical frameworks for the 'Anthropocene.' These have typically emphasized relationality, and in addition, the imbrication of human beings with non-human forms of life and the planet Earth itself. In some work we can see how 'posthuman' assumptions converge with feminist understandings of relationality.[47] A key theme in this work, for instance, has been the advocacy of a 'feminist ethics of care,' which places the emphasis on relations and includes 'such subjective feelings as compassion, love, and empathy.'[48] In these contributions, the de-centering of the human thus provides an alternative ontological point of departure from which to formulate novel ethical proposals to tackle climate change.

Finally, the rupture in IR brought about by the 'Anthropocene' has led to various forms of theoretical development and innovation to explain our current socio-ecological context. In response to the ontological and epistemological challenges posed, authors have been turning to novel approaches deemed more apt in accounting for the imbrication of the social and natural worlds. This work has been grouped, in a largely unsatisfactory manner, under the categories of 'posthumanism' or 'new-materialism.'[49] To be sure, the differences between what are essentially heterogeneous approaches can sometimes be quite profound; for instance, both Object-Oriented-Ontology (OOO) and Actor-Network Theory (ANT) may well seek to overcome the Cartesian subject-object dichotomy, yet the commitment to ontological relationality espoused is markedly different in both cases.[50] That said, it is also possible to identify some common ground between the various strands of what I will now refer to as 'posthumanism.'[51] For one thing, these approaches share a common disdain for an anthropocentrism and humanism which is understood to have led us to the brink of disaster in the first place. As Erika Cudworth and Stephen Hobden have pointed out, there is a generalized 'reaction against the view of human exceptionalism

---

[47] E. Allison (2017) 'Toward a Feminist Care Ethic for Climate Change,' *Journal of Feminist Studies in Religion*, 33(2), Fall 2017, p. 152.

[48] Ibid., p. 154.

[49] See the edited volume by C. Eroukhmanoff & M. Harker Eds. (2017) *Reflections on the Posthuman in International Relations*, E-International Relations Publishing. See also S. E. Wilmer & A. Žukauskaitė Eds. (2023) Life in the Posthuman Condition: Critical Responses to the Anthropocene, Edinburgh University Press.

[50] D. Rothe (2017) 'Global security in a posthuman age? IR and the Anthropocene challenge' in C. Eroukhmanoff & M. Harker Eds. *Reflections on the Posthuman in International Relations: The Anthropocene, Security and Ecology*, E-International Relations: E-Book.

[51] W. Connolly (2013) 'The 'New Materialism' and the Fragility of Things,' *Millennium: Journal of International Studies*, 41 (3), pp. 399–412; E. Cudworth & S. Hobden (2015) 'Liberation for Straw Dogs? Old Materialism, New Materialism, and the Challenge of Emancipatory Posthumanism,' *Globalizations*, 12(1) pp. 134–148.

(or anthropocentrism),'[52] which entails an erroneous understanding of humanity as that which is 'marked off from the huge diversity of non-human animal life' due to 'exceptional circumstances,' such as the possession of syntactical language or of 'free will.'[53] Human beings, it is argued, are not worthy of the Olympian position they have assigned themselves; that is, as all-seeing, all-knowing, and all-conquering subjects. Further, there is a common agreement on the radical interconnectivity of the world. Human beings are thus understood to be embedded in wider networks of relations, something which harbors quite profound implications for politics, security, and ethics in the 'Anthropocene.' For instance, a concern with 'posthuman security' has led Audra Mitchell to develop a 'worldly approach to security',[54] while Delf Rothe has focused on the 'visual' assemblages of environmental security composed of various human and non-human components.[55] The de-centering of the human subject, or state conceived as national community, typical in posthuman approaches would thus appear to offer a response to the ongoing deconstruction of the 'international' or 'global' catalyzed by notions of an 'Anthropocene.'

The 'Anthropocene,' Ontopolitics, and International Relations (IR)

While there is little doubt that the planet Earth is undergoing deeply troubling climatic changes, and that human beings have played a key role in altering the Earth System, the suitability of the concept 'Anthropocene' to capture this phenomenon has been open to debate. Critics have leveled at least two major charges against mainstream conceptions of the 'Anthropocene' found in environmental policy discourse, academia, and the mainstream media. First, some have questioned the narrative of an 'awakening' by climate scientists, who have purportedly just recently discovered the impact of human activity on the planet Earth.[56] This version of events is sometimes accompanied by a certain hubris which smuggles back in an anthropocentric overvaluation of science and its potential for rectifying current problems related to the climate and environment.[57] Furthermore, in failing to fully recognize the conceptual history and continuities of the 'Anthropocene' with its antecedents, there is a danger

---

[52] Cudworth & Hobden, ibid., p. 137.

[53] Ibid., p. 137.

[54] A. Mitchell (2014) 'Only Human? A Worldly Approach to Security,' *Security Dialogue*, 45(1), pp. 5–22.

[55] D. Rothe (2017) 'Seeing Like a Satellite: Remote Sensing and the Ontological Politics of Environmental Security,' *Security Dialogue*, 48(4), pp. 334–353.

[56] C. Bonneuil & J.B Fressoz (2016) *The Shock of the Anthropocene: the Earth, History and Us*, London: Verso, p. 11.

[57] Ibid.

that we might overlook the concept's inherent Eurocentrism and coloniality.[58] There is also ample room to fall into the trap of taking too literally the temporal characterization of a 'new' geological period. While attempting to pinpoint a starting point for the 'Anthropocene' is always likely to be politically fraught and contentious, stopping at the Industrial Revolution—in a sense logical in that it marks an important increase in human activity—[59]fails to situate environmental changes within a longer historicity of capitalist development and exploitation of the natural world.[60]

Second, the manifestly abstract and imprecise agent denoted in notions of an 'Anthropos,' or theories of 'human-induced' climate change, also turn out to be acutely problematic. As some authors working within Marxist or Feminist theoretical frameworks of IR have convincingly argued, it is simply not sufficient politically or ethically to designate 'human beings' per se as those responsible for the climate and ecological crisis. To avoid the depoliticizing tendencies of the concept 'Anthropocene,' it is thus necessary to acknowledge which sections of humanity bear the greatest responsibility for inducing climate and environmental change. Far from constituting a fixed, stable, neutral, and stand-alone geo-scientific meaning, one which merely identifies the constitutive role of human beings in disrupting the Earth System, it may be more fruitful to examine how the 'Anthropocene' has been integrated within specific security discourses.[61]

Some authors in IR have taken up the task of interrogating the multiple discourses of the 'Anthropocene.' For Cameron Harrington, there have been two different and 'competing characterizations of the new epoch' in this field: one which understands the 'Anthropocene' as 'rupture,' and the other which views it in terms of 'entanglement.'[62] Meanwhile, Elisa Randazzo and Hannah Richter differentiate a 'science-led, discontinuous-descriptive' from a 'continuous-ontological' conceptualization of the 'Anthropocene.'[63] Other authors have also sought to analyze 'Anthropocene' debates, and 'how they may reconfigure Earth as political space.'[64] Through attention to the 'ontological claims, analytical themes and political concerns around which IR

---

[58] M. Simpson (2020) 'The Anthropocene as Colonial Discourse,' *Environment and Planning D: Society and Space*, 38(1), pp. 53–71.

[59] P. Crutzen & W. Steffen (2003) 'How Long Have We Been in the Anthropocene Era?' *Climatic Change*, 61(3), p. 254.

[60] J. W. Moore (2017) 'The Capitalocene, Part 1: on the nature and origins of our ecological crisis,' *The Journal of Peasant Studies*, 44(3), pp. 594–630.

[61] D. Simangan (2020) 'Where is the Anthropocene? IR in a new Geological Epoch,' *International Affairs*, 96(1), pp. 211–224.

[62] C. Harrington (2020) 'A Quantum Anthropocene? International Relations Between Rupture and Entanglement,' in *Non-Human Nature in World Politics: Theory and Practice* Eds. J.C. Pereira & A. Saramago, pp. 53–72.

[63] E. Randazzo and H. Richter (2021) 'The Politics of the Anthropocene: Temporality, Ecology, and Indigeneity,' *International Political Sociology*, 15, pp. 293–312; E. Randazzo and H. Richter (2024) 'Political Action in Planetary Times: Extinction Activism, Anthropocene Ontopolitics, Indigenous Complexities,' *Political Geography*, 112, 103,107.

[64] E. Lövbrand, M. Mobjörk, & R. Soder (2020) 'The Anthropocene and the Geo-political Imagination: Re-writing Earth as Political Space,' *Earth System Governance*, 4, 100,051, pp. 1–8.

debates on the Anthropocene currently seem to circle,' Lövbrand et al. have identified three 'Anthropocene' discourses, which they label 'the endangered world,' 'the entangled world,' and 'the extractivist world.'[65] The authors conclude that, while they share certain elements, these discourses entail 'important differences in the interpretations' of the 'Anthropocene,' something which will have 'significant implications for the future of world politics.'[66]

The interrogation and critique of hegemonic 'Anthropocene' discourses has gone hand in hand with the elaboration of other, some have argued, more appropriate designations of our contemporary socio-ecological context. For some authors, this entails fully taking stock of the deleterious effects of global capitalism and its constituent logics on the biosphere.[67] The insatiable pursuit of surplus value and profit, leading to the appropriation and exploitation of 'cheap nature' and destructive forms of 'environment-making,' have been captured in the formulation of the neologism: the 'Capitalocene.'[68] There has indeed been a vast array of neologisms put forward to capture the impact that human beings have been having on the Earth System. To map the relations between humans and non-humans Donna Haraway has offered the term 'Chthulucene.'[69] Drawing on Haraway's work, Cara Dagett has explored the implications of a 'world-making' approach to global politics.[70] From within eco-feminist frameworks, authors have suggested the 'Androcene' as the most appropriate one for highlighting the constitutive role of patriarchy in generating the climate crisis.[71] Alternatively, the 'Technocene' shifts the central focus on to technology.[72] Meanwhile, others like Eric Swyngedouw and Henrik Erntson have preferred to eschew the 'Anthropocene' and alternatives, instead deploying the term 'Anthropo-obScene' to contest what they believe to be its depoliticizing effects.[73]

---

[65] Ibid., p. 2.

[66] Ibid., p. 5.

[67] J.W. Moore (2016) *Anthropocene or Capitalocene? Nature, History, and the Crisis of Capitalism*, PM Press.

[68] J. W. Moore (2017) 'The Capitalocene, Part 1: On the Nature and Origins of Our Ecological Crisis,' *The Journal of Peasant Studies*, 44(3), pp. 594–630. This is not the only way that authors have sought to attribute responsibility to global capitalism for affecting climate and environmental change. For two very different approaches see also F. Guattari (2000) *The Three Ecologies*, London: The Athlone Press, and J.B. Foster (2016) 'Marxism in the Anthropocene: Dialectical Rifts on the Left,' *International Critical Thought*, 6(3), pp. 393–421.

[69] D. J. Haraway (2016) *Staying with the Trouble: Making Kin in the Chtulucene*, Duke University Press.

[70] C. Daggett 'World-Viewing as World-Making: Feminist Technoscience, International Relations, and the Aesthetics of the Anthropocene,' in *Science, Technology, and Art in International Relations* Eds. J.P. Singh, Madeline Carr, & R. Marlin-Bennett, Routledge.

[71] J. Vuillerod (2021) 'The Anthropocene is an Androcene: Three Ecofeminist Narratives,' *Nouvelles Questions Féministes*, 40(2), pp. 18–34.

[72] O. López-Corona and G Magallanes-Guijón (2020) 'It is Not an Anthropocene: It is Really the Technocene: Names Matter in Decision Making under Planetary Crisis,' *Frontiers in Ecology and Evolution*, p. 8.

[73] E. Swyngedouw and H. Erntson (2018) 'Interrupting the Anthropo-obScene: Immuno-Biopolitics and Depoliticizing Ontologies in the Anthropocene,' *Theory, Culture & Society*, 35(6), pp. 3–30.

I would like to return now to the ontological register as it is here where some other important critical work on the 'Anthropocene' can be found. As noted previously, some authors have argued that the 'Anthropocene' brings about the deconstruction of Modernist binary categories such as culture/nature or human/non-human. In conjunction, this seemingly new geological epoch has also served to reinforce the ontological claims being made by some variants of 'post-humanism' or 'new materialism,' whose frameworks have been deployed with renewed vigor to grasp socio-ecological relationality and render the Modernist pursuit of 'progress' or 'happy ending' defunct.[74] Yet for other authors such as David Chandler these moves are deeply troubling since they can harbor undesirable political implications.[75] This is the realm of 'ontopolitics,' which we can think of as the often-unrecognized ontological assumptions which underpin political theories, practices, modes of thought, and governmental mechanisms.[76] As Chandler puts it, 'the ontopolitics of the Anthropocene privileges the 'is' of the world over the 'ought' of attempts to carve out a separate human space.'[77] It is about 'putting the nature of entangled being at the centre of politics rather than the designs or goals of the human as subject,' meaning that 'governing thereby seeks to adapt or respond to the world rather than seeking to control or direct it.'[78] This 'affirmation' of the complexity and contingency of the 'Anthropocene' is not a cause for celebration since the result can be an impoverished account of human subjectivity and agency, and ultimately depoliticization. In a context of ensuing climate and environmental crisis whereby meaningful political responses are required, this apparent curtailment of human capacities becomes acutely problematic.

We agree with Chandler on the importance of placing an analytical focus on ontopolitics for a greater appreciation of the ways that we understand and respond to the 'Anthropocene.' Just as the term itself must remain open to political struggle and critique, so must the ontopolitical assumptions which ground its hegemonic framings. For that, and with a view to re-politicizing the 'Anthropocene,' what shall be required is an 'ontopolitics of critique' as opposed to an 'ontopolitics of affirmation.'[79] Nonetheless, as we have noted in this introductory section, discourses of the 'Anthropocene' are not uniform, and authors have been advocating various alternative terms to grasp our contemporary socio-ecological context. For that, it seems more apt to speak of 'Anthropocenes,' as some authors have already been doing.[80] At any rate, we contend that any 'ontopolitics of critique' must extend toward the interrogation of the different ways of framing the 'Anthropocene' or alternative terms

---

[74] D. Chandler (2018) *Ontopolitics in the Anthropocene: An Introduction to Mapping, Sensing and Hacking*, Routledge; see also Randazzo and Richter, 'The Politics of the Anthropocene'.

[75] Ibid.

[76] See also W. E. Connolly (1995) *The Ethos of Pluralization*, Minneapolis: University of Minnesota Press, and D. Campbell (2005) 'Beyond Choice: The Onto-Politics of Critique,' *International Relations*, 19(1), pp. 127–134.

[77] Chandler, 'Ontopolitics in the Anthropocene,' p. 16.

[78] Ibid., pp. 20–21.

[79] Ibid., p. 210.

[80] T. Toivanen et al. (2017) 'The many Anthropocenes: A Transdisciplinary Challenge for the Anthropocene Research,' *The Anthropocene Review*, 4(3), pp. 183–198.

used in its place. This should not only provide a greater understanding of the ontopolitical assumptions grounding these different framings, but may also help to advance discussions on the critical potential of the 'Anthropocene' for the elaboration of better and more robust solutions to the current climatic and environmental crisis.

One of the unique features of this volume is how it brings together a group of chapters which display a range of different framings and conceptualizations of our current socio-ecological context, showing how they can be of use for the study of IR. These chapters, written by a group of scholars from diverse academic and cultural backgrounds, take as their shared point of departure the 'Anthropocene' rupture in IR. The authors build on some of the work hitherto mentioned in this Introduction by pursuing, in their own ways, new directions in research areas like governance, security, geopolitics, migration-mobility, social movements and ethics. While some authors show a preference for the concept 'Anthropocene,' the ontopolitical assumptions grounding its framings are not always the same. Meanwhile, others offer critical reflection on its hegemonic iterations, while elaborating and deploying various alternative terms. Indeed, perhaps two of the only points on which the authors assembled here are in close agreement are: (1) there is a rapidly changing human-induced ecological context, and (2) how we frame this phenomenon harbors profound political and ethical implications for the theory and practice of International Relations.

The principal aim of this volume is to explore the ontopolitics of different framings of the 'Anthropocene' and their implications for the field of IR. We thus adopt a self-reflexive approach, which is concerned with the relationship between the ontological commitments vis-à-vis the nature of the World that is being perceived, and the political and ethical implications which derive from these situated perspectives. It is in this way that the book seeks to stake out its place and differentiate itself from other edited volumes with a similar concern with the 'Anthropocene' and IR.[81] Through the pursuit of this 'ontopolitics of critique,' which stands in sharp contrast to the ontopolitics of closure inherent in some formulations of the 'Anthropocene,'[82] the volume explores answers to a series of questions: what ontological assumptions underpin the different framings of the 'Anthropocene' or alternative terms used to conceptualize our socio-ecological context? How are relations between the social and natural worlds articulated? What ontopolitical assumptions are made concerning the place of human subjectivity/agency in these different framings? What are the political implications in terms of policy prescription or practice? What kind of ethics do the ontopolitical assumptions make possible/occlude?[83]

We expect that this edited volume should interest a broad range of academics, researchers, educators, and advanced students working within the field of IR. Our contribution should particularly appeal to those who are engaged with questions

[81] J. C. Pereira and A. Saramago (2020) *Non-Human Nature in World Politics: Theory and Practice*, Springer; Chandler, Muller & Rothe 'International Relations in the Anthropocene'.

[82] Chandler, 'Ontopolitics in the Anthropocene,' p. 209.

[83] These questions were partially mapped by Lövbrand et al. 'the Anthropocene and the geo-political imagination'.

related to climate and environmental change, and who are grappling with the implications of the 'Anthropocene' for IR. That is, the political and ethical implications of how this 'condition' is framed.[84] In a more general sense, we hope that this collection of essays will attract the attention of a broader audience grappling with the concept 'Anthropocene,' and likewise enrich ongoing debates over how best to conceive of our contemporary ecological context.

In the second chapter, Joana Castro Pereira, João Terrenas, and Inês Ferreira de Sousa show how the geographical space of 'the Amazon' constitutes an important object of study through which to advance discussions in International Relations (IR) within the broader context of the 'Anthropocene.' The authors thus seek to show how this regional focus can generate novel ontological, analytical, and ethico-political perspectives which could be of use to IR. These include, 'relational, multicultural approaches to land governance and the bioeconomy,' the 'politics of freshwater ecosystems and hydrological connectivity,' and the 'interspecific character and implications of environmental militarisation.' However, this is not simply a case of mining indigenous ontological and epistemological resources. Rather, the authors emphasize the need for dialogue between different world-views, thereby allowing for deeper reflection on how we can grasp the range of challenges and opportunities vis-à-vis the conservation of the Amazon.

The chapter written by Ana Flávia Barros-Platiau et al. takes as its principal concern security in the 'Anthropocene.' For the authors, the ushering in of this new geological epoch brings with it unique challenges due to the intertwinement of two major clusters of security threats: on the one hand, there are those which emanate from climate and environmental degradation, while on the other, there are those which stem from the vulnerabilities associated with cyber-space. As the authors point out, the same technologies which are becoming indispensable for the management of climate change 'can also be employed as weapons to destabilize societies and promote violence or panic.' Since climate and cybersecurity threats are interconnected, there is thus a need for new global governance arrangements and new forms of diplomacy in the 'Anthropocene.' Ultimately, the chapter seeks to spark further discussion on how we might negotiate the interconnectedness of security threats in the 'Anthropocene,' and moreover in a period of heightening geopolitical tensions.

In Gabriela Kuetting's chapter, the concept 'Anthropocene' is taken as a point of departure for an exploration of the theme of multi-species justice. First, the author notes how the commodification of other-than-human species constitutes a vital part of the 'Anthropocene,' meaning that debates surrounding the question of multi-species do not merely take on ethical import. This spurs a survey of how this topic has been tackled in existing literature, exploring diverse contributions from Indigenous to Environmental Justice scholarship. Kuetting concludes that while 'our current understanding of nature-society relations lacks recognition for other species as 'worthy'

[84] The term 'condition' is used by D. Chandler, F. Müller and D. Rothe (2021) *International Relations in the Anthropocene: New Agendas, New Agencies and New Approaches*, Palgrave Macmillan, p. 3.

of representation,' a rights-based approach remains 'the most promising avenue' for achieving this.

Yoan Molinero-Gerbeau and Victor Pérez Segura's chapter is primarily conceptual in its orientation, seeking to interrogate Thomas Nail's elaboration of the 'Kinocene' as a way of grasping our contemporary socio-ecological context. While the authors agree on the utility of the concept in accounting for the 'increasing and intensifying mobilities' of our contemporary context, they prefer to situate this within a broader 'World-ecology' or 'Capitalocene' framework. To show this, the authors explore the topic of migration. Through an examination of key categories 'environmental migrants' and 'climate refugees,' the chapter aims to show how mobilities in the Capitalocene are affected by two dualisms: society/culture-nature and legitimate/illegitimate. The argument advanced is that 'this framework for understanding mobility makes it possible to establish a favourable regime for capitalist development, in which the high mobility of capital and capitalists contrasts with the restricted mobility of labour.'

In the following chapter, Álvaro San Román analyses two of the principal 'metanarratives' which give meaning to our present socio-ecological context: the 'Anthropocene' and the 'Capitalocene.' Bringing them into dialogue with each other, the author seeks to draw out the shortcomings of both, which seemingly derives from their inability to avoid succumbing to anthropocentrism. Yet the primary concern in this chapter is with the constitutive role that technology has played in shaping our contemporary socio-ecological context. Grasping 'technology' not simply in its material form, but rather as part of a Western 'worldview' or 'hegemonic cultural system', the author argues that the most appropriate term for our contemporary epoch is the 'Techno-Westernocene.' In this sense, San Román seeks to build on the work of previous authors who have been developing concepts like the 'Technosphere' or 'Technocene.'

In his chapter Matthew Robson carries out a critique of the 'Capitalocene' as an increasingly prominent framing for our contemporary epoch. Two perceived limitations of its elaboration in the academic literature are identified and addressed: on the one hand, Robson points to an insufficient theorization of the relationship between global capitalism and states in the Capitalocene; on the other, he claims that there is an inadequate elaboration of how change might take place, and the role that human beings might play in these processes. The author turns to Marxist IR scholarship and the assemblage work of Gilles Deleuze and Felix Guattari for insights into how these questions might be answered. The chapter first aims to show how a Deleuzoguattarian assemblage approach allows one to grasp the dynamic non-deterministic relationship between global capitalism and its state mechanisms or 'models of realization,' and also how both work in tandem to disrupt the geological strata on which they are dependent. Second, a focus on different kinds of assemblage can seemingly generate insights into how bottom-up contestation and resistance to capitalist states may emerge to mitigate climate change. Ultimately, Robson argues that the assemblage work of Deleuze and Guattari offers considerable, and relatively underexplored, theoretical resources for approaching the 'Capitalocene,' whether we retain this term or not.

In the final chapter, Matthew Robson summarizes some of the main arguments developed, synthesizing the contributions in line with the broader aims of the volume. Highlighting once more the variety of ways that the contributing authors have conceptualized the contemporary socio-ecological context, Robson aims to carry out an 'ontopolitics of critique' of the preceeding chapters. As mentioned previously, this entails interrogating the 'often-unrecognized ontological assumptions' which underpin the authors' framings of the 'anthropocene'. This final conclusions chapter thus reiterates the twofold contribution made by the volume to the field of IR: 1) in showcasing the diversity of conceptualizations of our socio-ecological context and how they can be used to study some of the key topics in the field, and 2) in showing how these framings are invariably grounded on ontopolitical assumptions that are contestable, and which must therefore remain open to critique.

Chapter 2
International Relations and the Amazon in the Anthropocene

Joana Castro Pereira, João Terrenas, and Inês Ferreira de Sousa

Introduction

Over the past decade, the Anthropocene has emerged as one of most important topics for the field of International Relations (IR). This is reflected in the growing number of articles, books, special issues, and edited volumes that have been published on this topic. At the same time, there has been a trend towards developing research groups, workshops, seminars, courses, and international academic conferences that aim to place the Anthropocene at the heart of disciplinary debate. These developments are, to some extent, normal. The planetary challenges raised by this proposed new geological epoch directly intersect with the research agenda of a discipline that is first and foremost concerned with matters of global security and transnational cooperation. In light of these debates, it is evident that IR is in a privileged position to help us make better sense of the Anthropocene and tackle the wicked problems that it poses to earthly beings. IR's frameworks for analysis, if correctly adjusted, can be important tools to understand the transnational systems and processes that led to the emergence of the Anthropocene, from colonialism to capitalism or militarism. In addition to that, and given its foundational concern with informing practice and policy-making, the discipline can help provide relevant insights into the collective pathways needed to transform socio-ecological relations and address the challenges posed by the Anthropocene to humanity.

J. C. Pereira (✉)
University of Porto, Porto, Portugal
e-mail: jcpereira@letras.up.pt

J. Terrenas
University of Beira Interior, Covilhã, Portugal
e-mail: joao.terrenas@ubi.pt

I. F. de Sousa
NOVA University of Lisbon, Lisbon, Portugal
e-mail: inesafsousa@gmail.com

M. Robson (ed.), *The Anthropocene, Ontopolitics, and International Relations*,
https://doi.org/10.1007/978-981-96-2925-1_2

In this context, it would only be natural to see the Amazon emerging as a major topic of concern in debates about IR in the Anthropocene. This region provides evidence of the complex challenges intrinsic to environmental governance and planetary politics. By foregrounding the political significance of non-human beings and more-than-human networks, it also questions our most basic assumptions about international relations. It forces us to think politics and ethics beyond the human, to search for ways of knowing and being with rather than merely in the world. Crucially, the Amazon is one of the most important spaces for ensuring the long-term survival of humanity and the sustainability of planetary life. With an area of around 5.5 million km^2, the Amazonian borders cut across nine different countries. The region is home to numerous ethnic groups and Indigenous cultures, while its rivers and lands harbour unique species of plants and animals.[1] The Amazon has also been the focus of recurrent academic and political debates on issues such as conservation, sovereignty, development, or security, whereas its stability has been directly linked to the stability of the Earth system as a whole. Simply, the survival of this biome is deeply intertwined with the future of life on Earth.

Yet, as with many other major biomes, the region has thus far remained largely invisible to the gaze of IR scholars. The scant literature engaging with the Amazon from an IR perspective has focused primarily on researching the global, regional, and Indigenist politics of forest and water governance,[2] or in exploring (socio)environmental activism and advocacy within and beyond the region.[3] Taking a different approach, Picq[4] used the Amazon to criticise the inequality in knowledge production within IR, showing the region's role in the historical constitution of international relations and therefore the global character of Amazonian experiences that

[1] SPA, *Amazon Assessment Report 2021* (New York: Science Panel for the Amazon, 2021).

[2] A. F. Barros-Platiau, N. Søndergaard, and J, Prantl, "Policy Networks in Global Environmental Governance: Connecting the Blue Amazon to Antarctica and the Biodiversity Beyond National Jurisdiction (BBNJ) Agendas", *Revista Brasileira de Política Internacional* 62, no. 2 (2019): e006; J. C. Pereira and E. Viola, *Climate Change and Biodiversity Governance in the Amazon: At the Edge of Ecological Collapse?* (New York: Routledge, 2022); L. de Oliveira Paes, "The Amazon Rainforest and the Global–Regional Politics of Ecosystem Governance", *International Affairs* 98, no. 6 (2022): 2077–2097; V. K. Gonçalves, "Carbon Offset from the Amazon Forest to Compensate Aviation Emissions: Global Solution, Local Struggles", *Earth System Governance* 14 (2022): 100160; V. M. Chase, "The Changing Face of Environmental Governance in the Brazilian Amazon: Indigenous and Traditional Peoples Promoting Norm Diffusion", *Revista Brasileira de Política Internacional* 62, no. 2 (2019): e008.

[3] C. Y. A. Inoue and M. Franchini, "Socio-environmentalism", in *International Relations from the Global South: Worlds of Difference*, ed. A. B. Tickner and K. Smith (New York: Routledge, 2020), 296–314; K. Hochstettler and M. E. Keck, *Greening Brazil: Environmental Activism in State and Society* (Durham: Duke University Press, 2007); P. F. Moreira et al., "South–South Transnational Advocacy: Mobilizing Against Brazilian Dams in the Peruvian Amazon", *Global Environmental Politics* 19, no. 1 (2019): 77–98.

[4] M. Picq, "Rethinking IR from the Amazon", *Revista Brasileira de Política Internacional* 59, no. 2 (2016): e003.

scholars in the field have overlooked. More recently, Inoue,[5] Pereira and Gebara,[6] and Pereira and Terrenas[7] have linked the Amazon to the Anthropocene and IR theory by bringing to the fore Indigenous ways of knowing and being, thus helping advance the emergent relational and pluriversal agendas within the field.[8]

This literature has shown that delving into the complexities and experiences of the Amazon can help scholars reflect upon IR theories' own limits and possibilities in the Anthropocene. However, a more substantive understanding of how, in the face of a major planetary crisis, engaging with the Amazon can make IR better prepared to understand, problematise, and contribute to address the challenges raised by the Anthropocene is still lacking. In this chapter, we address this question by taking the Amazon as an entry point to advance a research agenda that demonstrates the analytical, normative, and transformative possibilities that are already immanent within the region.

The argument starts by situating the Anthropocene within IR and related fields, contextualising the growing significance of this topic for thinking world politics otherwise and exploring the different ways in which scholars have responded to it. Following from that, it looks into the Amazon and identifies some issues wherein disciplinary discussions could strongly benefit from a more systematic engagement with the region. We argue that the Amazon brings to the fore at least four broad aspects of planetary politics that can enrich extant debates on the Anthropocene and strengthen the analytical and ethico-political potential of IR regarding the transformative potential of relational, socio-cultural approaches to land governance and the bioeconomy; regional conceptions and experiences of more-than-human personhood and community formation; the transnational politics of freshwater ecosystems and hydrological connectivity; and the interspecific risks and threats of environmental militarisation. We demonstrate how, despite its overwhelming absence from most debates about world politics, the Amazon holds the potential to help consolidate the growing effort to develop "post-anthropocentric IR".[9]

Still, it is not simply a matter of bringing the Amazon into IR. This move should be complemented by an effort to also bring IR more systematically into the region. In the concluding section of this chapter, we briefly explore how doing so also allows us to identify alternative approaches that can help protect this biome and reduce the

---

[5] C. Y. A. Inoue, "Worlding the Study of Global Environmental Politics in the Anthropocene: Indigenous Voices from the Amazon", *Global Environmental Politics* 18, no. 4 (2018): 25–42.

[6] J. C. Pereira and M. F. Gebara, "Where the Material and the Symbolic Intertwine: Making Sense of the Amazon in the Anthropocene", *Review of International Studies* 49, no. 2 (2023): 319–338.

[7] J. C. Pereira and J. Terrenas, "Towards a Transformative Governance of the Amazon", *Global Policy* 13, no. S3 (2022): 60–75.

[8] M. Kurki, "Relational Revolution and Relationality in IR: New Conversations", *Review of International Studies* 48, no. 5 (2021): 821–836; T. Trownsell, N. C. Behera, and G. Shani, "Introduction to the Special Issue: Pluriversal Relationality", *Review of International Studies* 48, no. 5 (2022): 787–800.

[9] J. C. Pereira and A. Saramago, "Introduction: Embracing Non-Human Nature in World Politics", in *Non-Human Nature in World Politics: Theory and Practice*, ed. J C. Pereira and A. Saramago (Cham: Springer, 2020), 1–9.

threats and vulnerabilities faced by its multiple socio-ecological communities. While it is critical to recognise how IR can learn and evolve by engaging with the Amazon, we maintain that the region should not simply be viewed as a site for extracting concepts, practices, or forms of knowledge to make the discipline more vibrant and attuned to the Anthropocene. Instead, our current understanding of the socio-ecological challenges and problems faced by its more-than-human communities can also benefit from applying IR's theoretical frameworks to the transnational political configurations that, over the centuries, have made and remade this region. To close the gap between IR and the Amazon, future research should also point in this direction.

The Anthropocene in IR (and Beyond)

Since it was first coined by Crutzen and Stoermer,[10] the concept of the Anthropocene, a proposed new geological epoch characterised by the deep and destructive way in which human activities are shaping the planet and its future, has rapidly made its way into the natural, social, and human sciences. This also holds true for IR. However, it rapidly became apparent how unprepared the field is in the face of the planet's new socio-ecological conditions, both institutionally and intellectually.[11] By placing the state as the means and end of international security, by separating nature from society, or by upholding linearity over complexity, the conventional thinking tools of IR build images that are not only inadequate and incomplete, but also dangerous in the Anthropocene.[12] As some have pointed out, this has impeded the discipline's ability to engage with and contribute to environmental debates.[13]

The new planetary real exposed the complex interconnectedness between multiple and seemingly disparate policy areas and crises. It also blurred the distinction between humans and non-humans and, in doing so, revealed the mutually constitutive relations among nature, culture, and society.[14] Challenging the belief that human endeavours

---

[10] P. J. Crutzen and E. F. Stoermer, "The 'Anthropocene'", *Global Change Newsletter* 41 (2000): 17–18.

[11] A. Burke et al., "Planet Politics: A Manifesto from the End of IR," *Millennium: Journal of International Studies* 44, no. 3 (2016): 499–523.

[12] J. C. Pereira, "Towards a Politics for the Earth: Rethinking IR in the Anthropocene", in *International Relations in the Anthropocene: New Agendas, New Agencies and New Approaches*, ed. D. Chandler, F. Müller, and D. Rothe (Cham: Palgrave Macmillan, 2021), 21–37.

[13] C. Harrington, "The Ends of the World: International Relations and the Anthropocene", *Millennium: Journal of International Studies* 44, no. 3 (2016): 478–498; Pereira, "Towards a Politics for the Earth".

[14] Burke et al., "Planet Politics"; A. Burke and S. Fishel, "Power, World Politics, and Thing-Systems in the Anthropocene", in *Anthropocene Encounters: New Directions in Green Political Thinking*, ed. F. Biermann and E. Lövbrand (Cambridge: Cambridge University Press, 2019), 87–108; Harrington, "The Ends of the World"; D. Rothe, F. Müller, and D. Chandler, "Introduction: International Relations in the Anthropocene", in *International Relations in the Anthropocene*, ed. Chandler, Müller, and Rothe, 1–6; Pereira, "Towards a Politics for the Earth"; J. C. Pereira and J.

are separate from the environment in which they take place,[15] the Anthropocene not only questions IR's conventional methodologies, categories, and institutions, but also demands their rewriting.[16] In this way, scholars have been advocating for a transformation of the discipline, challenging its state-centric worldviews, positivist and rationalist underpinnings, and anthropocentric ethics.[17] Doing so, they uphold, is a precondition so that IR can better address the reality of a planet whose liveability is increasingly threatened by deep environmental transformations caused by human induced stresses in ecosystems.[18]

As defended by Burke et al.,[19] the Anthropocene invites (or rather forces) IR to question how it understands world politics in a manner that, at least until now, the discipline has not been able and willing to. Without reassessing its concepts and assumptions in light of our current predicament, IR risks becoming unable to help us think and move through these multiple and interweaving planetary crises. The Anthropocene forces us to cross boundaries, to approach the planet from various angles, to search for alternatives across multiple scales, and to think about politics and ethics beyond our own species. In breaking the ontological divide between nature and society, these crises also push us to question the epistemological division between the natural sciences, social sciences, and humanities.[20] Interdisciplinarity is therefore essential so that IR can make sense of the complexity of ecological crises.[21] It is along these lines that Burke et al.[22] advocate for "Planet Politics" as a new paradigm for research—an agenda aimed at cutting across different levels of analysis and exploring alternative forms of both governance and subversion, regulation and resistance, in search for strategies of mediating socio-ecological relations that no longer replicate the anthropocentric assumptions of the Holocene.

Others have argued that the Anthropocene requires policy institutions to abandon liberal, top-down, command-and-control, modernist methods of knowing and governing the world and, instead, invest in the development of bottom-up, inclusive approaches to governance through social change.[23] In this new epoch, politics

---

Renner, "Animals in International Relations: A New Research Agenda", *International Relations* 37, no. 3 (2023): 389–397; Pereira and Gebara, "Where the Material and the Symbolic Intertwine".

[15] S. Dalby, *Anthropocene Geopolitics: Globalization, Security, Sustainability* (Ontario: University of Ottawa Press, 2020); S. Dalby, "Rethinking Geopolitics: Climate Security in the Anthropocene", *Global Policy* 5, no. 1 (2014): 1–9.

[16] Burke et al., "Planet Politics".

[17] Pereira, "Towards a Politics for the Earth".

[18] IPCC, *Climate Change 2023: Synthesis Report. Contribution of Working Groups I, II and III to the Sixth Assessment Report of the Intergovernmental Panel on Climate Change* (Geneva: Intergovernmental Panel on Climate Change, 2023).

[19] Burke et al., "Planet Politics".

[20] Pereira, "Towards a Politics for the Earth".

[21] Harrington, "The Ends of the World"; Pereira, "Towards a Politics for the Earth".

[22] Burke et al., "Planet Politics", 507.

[23] D. Chandler, E. Cudworth, and S. Hobden, "Anthropocene, Capitalocene and Liberal Cosmopolitan IR: A Response to Burke et al.'s 'Planet Politics'", *Millennium: Journal of International Studies* 46, no. 2 (2017): 190–208.

and governance should not ignore the capitalist roots of the Earth crisis, as there are no real solutions under the very system that enabled the current crisis to unfold.[24] This recognition will allow IR to address the varying implications of capitalism upon different societies and locate the responsibilities that existing political and economic systems hold in planetary destruction.[25] Some authors have indeed begun to counter a tendency to generalise humanity's role in the current ecological crisis by incorporating feminist and decolonial perspectives into IR's engagement with the Anthropocene.[26] It is often argued that the term Anthropocene refers to humanity as a single and homogeneous force, thus overlooking the gender, class, and racialised power relations that made socio-ecological crises possible in the first place. As such, scholarship started to favour other terms to characterise the emerging epoch, such as Capitalocene—placing environmental degradation in relation to capitalist economies and the history of domination[27]—or Plantationocene—which suggests that the current ecological crisis is a consequence of historical plantation agriculture, focusing on the ideas of control and modernisation of nature[28]—and Eurocene—which attributes responsibility to European ways of life, its economies, its hierarchies, colonialism, and exploitation.[29]

Regardless of how it is characterised, scholars agree that dominant IR theories and current institutional frameworks have been largely blind to the condition of the Anthropocene, thus failing to address the destructiveness that it carries.[30] Existing institutions, designed for the stable conditions of the Holocene, lack the tools to address the uncertainty and non-linearity of the new epoch, which are exacerbated by the existence of thresholds and tipping points where a relatively small disturbance can cause a significant, disproportional change in a system and trigger large scale irreversible environmental changes.[31] As such, it is argued that present institutions

---

[24] Chandler, Cudworth, and Hobden, "Anthropocene, Capitalocene and Liberal Cosmopolitan IR"; M. J. Albert, "Capitalism and Earth System Governance: An Ecological Marxist Approach", *Global Environmental Politics* 20, no. 2 (2020): 37–56.

[25] E. Lövbrand, M. Mobjörk, and R. Söder, "The Anthropocene and the Geo-Political Imagination: Re-Writing Earth as Political Space", *Earth System Governance* 4 (2020): 100051.

[26] C. McEwan, "Decolonizing the Anthropocene," in *International Relations in the Anthropocene*, ed. Chandler, Müller, and Rothe, 77–94; D. Haraway, "Anthropocene, Capitalocene, Plantationocene, Chthulucene: Making Kin", *Environmental Humanities* 6, no. 1 (2015): 159–165; F. Müller, "Agency in More-than-Human, Queerfeminist and Decolonial Perspectives", in *International Relations in the Anthropocene*, ed. Chandler, Müller, and Rothe, 251–269.

[27] Haraway, "Anthropocene, Capitalocene, Plantationocene, Chthulucene"; J. W. Moore, *Capitalism in the Web of Life: Ecology and the Accumulation of Capital* (London: Verso, 2015).

[28] J. Davis et al., "Anthropocene, Capitalocene, … Plantationocene?: A Manifesto for Ecological Justice in an Age of Global Crises," *Geography Compass* 13, no. 5 (2019): e12438; Haraway, "Anthropocene, Capitalocene, Plantationocene, Chthulucene".

[29] J. V. Grove, *Savage Ecology: War and Geopolitics at the End of the World* (Durham: Duke University Press, 2019).

[30] Burke et al., "Planet Politics"; Pereira, "Towards a Politics for the Earth"; Rothe, Müller, and Chandler, "Introduction".

[31] P. Wassmann and T. M. Lenton, "Arctic Tipping Points in an Earth System Perspective," *Ambio* 41, no. 1 (2012): 1–9.

seem to be incapable of anticipating and acting upon emerging threats, as they were designed to "address structured, short-term, direct cause and effect issues, providing simple solutions that produce immediate effects".[32] For Dryzek and Pickering,[33] these institutions suffer from a "pathological path dependency" where they suppress information concerning the Earth system to prioritise economic growth. The authors propose that Anthropocene political institutions and governance practices should be based on the notion of "ecological reflexivity", where the cyclical recognition of, reflection on, and response to changes of socio-ecological systems enable us to respond to early warnings of ecological state shifts and thus succeed in an unstable and unpredictable environment. They underline the need for institutions to, when needed, be able "to be something different, rather than just do different things".[34] This, they argue, sets them apart from adaptive governance practices. Additionally, the authors recognise the need for those institutions to listen to the non-human world, so as to address the well-being of socio-ecological systems, rather than solely social systems.

Other scholars have also begun reflecting on what governance should look like in the Anthropocene and propose a range of approaches.[35] Namely, an increasing number of scholars advocate for the implementation of transformative governance models to offer suitable responses to the challenges presented by the Anthropocene.[36] It is held that transformative governance can bring about transformative change to socio-ecological systems, which is necessary to address both environmental destruction and structural inequalities, as well as to promote sustainability across scales, issues, places, and sectors. Hence, it puts forward an alternative to top-down, managerial, technocratic, fragmented, and anthropocentric governance approaches, which fall short in addressing the complexity and uncertainty of the Anthropocene and, on the contrary, add to the destruction of systems.[37] Transformative governance aims, instead, to be an integrative, inclusive, adaptive, and pluralist approach that through political, ethical value-driven, and co-decision-making processes, including the participation of those whose voices are currently not being heard, can address

[32] J. C. Pereira and E. Viola, "Catastrophic Climate Change and Forest Tipping Points: Blind Spots in International Politics and Policy", *Global Policy* 9, no. 4 (2018): 520.

[33] J. S. Dryzek and J. Pickering, *The Politics of the Anthropocene* (Oxford: Oxford University Press, 2019).

[34] Dryzek and Pickering, *The Politics of the Anthropocene*, 17.

[35] B. Bornemann, "Environmental Governance in the Anthropocene: Challenges, Approaches and Critical Perspectives", in *International Relations in the Anthropocene*, ed. Chandler, Müller, and Rothe, 311–329.

[36] I. J. Visseren-Hamakers and M. Kok, *Transforming Biodiversity Governance* (Cambridge: Cambridge University Press, 2022); Pereira and Terrenas, "Towards a Transformative Governance of the Amazon"; U. Pascual et al., "Governing for Transformative Change across the Biodiversity–Climate–Society Nexus", *BioScience* 72, no. 7 (2022): 684–704.

[37] Pereira and Terrenas, "Towards a Transformative Governance of the Amazon"; Visseren-Hamakers and Kok, *Transforming Biodiversity Governance.*

contemporary socio-ecological issues.[38] The participation of typically marginalised communities, whether they are human or non-human, in governance practices and decision-making processes is argued to be an integral part of designing democratic institutions in the Anthropocene.[39] As Mert[40] advises, democracy needs to be rethought in a new scale to meet the requirements of contemporary societies and challenges, looking to counter the technocratic and authoritarian tendencies that are frequently proposed as possible solutions to the ecological crisis. To address many of the democratic challenges the Anthropocene poses, we need to deconstruct existing practices and traditions, build reflexive global institutions, which can be "adaptable under swiftly changing circumstances",[41] commit to transdisciplinary knowledge co-production, with the contribution of different stakeholders, and embark on democratic experimentation to "invent ways to include non-human actants".[42]

The Anthropocene also sheds a new light on some of IR's fundamental issues, such as security, sovereignty, and power, while exhausting their traditional conceptions. With environmental concerns, particularly climate change, increasingly being analysed through a security lens, scholars have warned against using traditional, state-centric, and fragmented conceptions of security, which overlook the impact that our economic systems and ways of life have in causing insecurities in the Anthropocene.[43] For Dalby,[44] security should mean reshaping our economies to be flexible and free from fossil fuel dependence, which is essential to avoid "old-fashioned geopolitics".[45] Scholars have also begun considering how the consequences of the exploitation of nature by humans, which led to the Anthropocene, cause not only physical insecurities to communities but also ontological insecurities, in the face of unprecedented existential threats (Mattos & Henao 2021; Farbotko 2019).[46] This

---

[38] Pascual et al., "Governing for Transformative Change"; Visseren-Hamakers and Kok, *Transforming Biodiversity Governance*.

[39] A. Burke and S. Fishel, "Across Species and Borders: Political Representation, Ecological Democracy and the Non-Human", in *Non-Human Nature in World Politics*, ed. Pereira and Saramago, 33–52; A. Mert, "Challenges to Democracy in the Anthropocene", in *International Relations in the Anthropocene*, ed. Chandler, Müller, and Rothe, 291–309.

[40] Mert, "Challenges to Democracy in the Anthropocene".

[41] A. Mert, "Democracy in the Anthropocene: A New Scale", in *Anthropocene Encounters*, ed. Biermann and Lövbrand, 144.

[42] Mert, "Challenges to Democracy in the Anthropocene", 303.

[43] Dalby, *Anthropocene Geopolitics*; Dalby, "Rethinking Geopolitics".

[44] S. Dalby, "Environmental Security and the Geopolitics of the Anthropocene", in *International Relations in the Anthropocene*, ed. Chandler, Müller, and Rothe, 137–153; S. Dalby, "To Build a Better World: Securing Global Life after Fossil Fuels," in *Anthropocene (In)Securities: Reflections on Collective Survival 50 Years after the Stockholm Conference*, ed. E. Lövbrand and M. Mobjörk (Oxford: Oxford University Press, 2021), 84–98.

[45] Dalby, "Rethinking Geopolitics", 8.

[46] B. R. B. Mattos and S. G. Henao, "Whose Security/Security for Whom? Rethinking the Anthropocene through Ontological Security", in *Anthropocene (In)Securities*, ed. Lövbrand and Mobjörk, 117–130; C. Farbotko, "Climate Change Displacement: Towards Ontological Security", in *Dealing with Climate Change on Small Islands Towards Effective and Sustainable Adaptation?*, ed. C. Klöck and M. Fink (Göttingen: Göttingen University Press, 2019), 251–266.

has been illustrated through the study of the effects of climate change as an ontological threat in small island states and its communities, which in turn revealed the complexity of what it means to be secure in the Anthropocene.[47] Scholars claim that, given its characteristics, the new epoch requires us to abandon the assumptions that a threat emanates from the "outside",[48] that there is a clear distinction between referent object and threat, and that a boundary exists between nature and society in our security discourses.[49] The blurring of this boundary also exposes the inadequacy of anthropocentric approaches to security, prompting scholars to explore alternative frameworks that recognise the interconnectedness and shared vulnerability among all living beings.

McDonald[50] offers the framework of ecological security, recognising that, in the context of the Anthropocene, emphasis should be placed on ensuring ecosystem resilience through addressing the factors that render them insecure. Apart from disclosing the frames of national, international, and human security as inadequate for the current context, he defends that placing the focus on ecosystem security can allows us to prioritise the rights and needs of vulnerable communities, both human and non-human, both present and future. Leep[51] also explores how one might incorporate the non-human more explicitly in security discourses. Specifically, his argument focuses on individual security and the notion of personhood, contesting former frameworks that have excluded non-human individuals from being classified as a "person" and thus who is able to experience security. In turn, Harrington[52] explores a vision of security for the Anthropocene that builds upon the ethics of care—a view that, in taking the world and its beings as relational, foregrounds threats arising from the entanglement among humans, non-humans, and things. This, he argues, would allow us to study flows of violence which are usually absent from IR analyses, such as structural violence, micro-practices of violence, or violence against non-human forms of life, as well as to overcome the scale and temporal restrictions that conventional approaches to security studies cannot capture. Similarly, Mitchell[53] proposes the concept of "mundicide" to describe harm inflicted to worlds co-constituted by diverse beings, both human and non-human, and to the "condition

[47] Mattos and Henao, "Whose Security/Security for Whom?"; Farbotko, "Climate Change Displacement".

[48] S. Hamilton, "Securing Ourselves from Ourselves? The Paradox of 'Entanglement' in the Anthropocene", *Crime, Law and Social Change* 68, no. 5 (2017): 579–595.

[49] M. Fagan, "Security in the Anthropocene: Environment, Ecology, Escape", *European Journal of International Relations* 23, no. 2 (2017): 292–314.

[50] Matt McDonald, *Ecological Security: Climate Change and the Construction of Security* (Cambridge: Cambridge University Press, 2021).

[51] M. Leep, "Introduction to the Special Issue: Multispecies Security and Personhood", *Review of International Studies* 49, no. 2 (2023): 181–200.

[52] C. Harrington, "Posthuman Security and Care in the Anthropocene", in *Reflections on the Posthuman in International Relations: The Anthropocene, Security and Ecology*, ed. C. Eroukhmanoff and M. Harker (Bristol: E-International Relations Publishing, 2017), 73–86.

[53] A. Mitchell, "Only Human? A Worldly Approach to Security", *Security Dialogue* 45, no. 1 (2014): 5–21.

of worldliness", thus enabling them to be subjects of in/security. Importantly, Lappe-Osthege and Duffy[54] discuss how individuals involved in wildlife crime exploit their understanding of avian behaviour to engage in illegal activities, advocating for incorporating knowledge of animal cultures into conservation policies to enhance global environmental governance and protect animal species. While different, all of these approaches respond to the call to think IR relationally, bringing our focus to processes and relations rather than isolated things and beings, which in turn help uncover the relational entanglements and more-than-human relations that the discipline has long overlooked.[55]

Indeed, studying more-than-human relations becomes a priority given the challenge that the emerging epoch poses on the dichotomies—human/nature, subject/object, agent/structure, and inside/outside—that the discipline's dominant ontologies take for granted.[56] There has been an increasing effort to move past anthropocentric notions of power and sovereignty that overlook more-than-human interactions, and which can originate dangerous practices and institutions. Burke and Fishel[57] propose an alternative theory that comprehends the complexity of power in "thing-systems" and admits the connection between society and nature. According to them, the Anthropocene undermined traditional, ecomodernist, human-centric conceptions of power predicated upon humanity's capacity to shape, manage, control, or even transform the environment in which it exists. As power cuts across a complex entanglement of more-than-human systems, the traditional Hobbesian notion of sovereignty, according to which only a state can hold the monopoly of power over a given territory, is also inadequate for the Anthropocene.[58] Traditional concepts of sovereignty are deeply anthropocentric and thus stand as a limitation to the recognition of non-human forms of sovereignty, namely that of animals and wild animal communities, thus hindering the ability for interspecies and intercommunity interactions and relations.[59] Critics, however, warn against granting rights and sovereignty to wild animal communities as it depends on human power and caution that real change should come through changing human concepts and institutions.[60] For instance, Youatt[61] uses the case of wild grey wolves in Isle Royale to explore the potential of animal sovereignties. Through the analysis of this and other cases, he demonstrates how IR concepts—such as borders, security, and sovereignty—are, in fact, the result of

---

[54] T. Lappe-Osthege and R. Duffy, "International Relations and the Non-Human: Exploring Animal Culture for Global Environmental Governance", *Review of International Studies* (early view).

[55] M. Kurki, *International Relations in a Relational Universe* (Oxford: Oxford University Press, 2020).

[56] C. Harrington, "A Quantum Anthropocene? International Relations Between Rupture and Entanglement", in *Non-Human Nature in World Politics*, ed. Pereira and Saramago, 53–72.

[57] Burke and Fishel, "Power, World Politics, and Thing-Systems in the Anthropocene".

[58] Burke and Fishel, "Power, World Politics, and Thing-Systems in the Anthropocene".

[59] S. Donaldson and W. Kymlicka, *Zoopolis: A Political Theory of Animal Rights* (Oxford: Oxford University Press, 2011).

[60] D. Wadiwel, *The War Against Animals* (Boston: Brill, 2015).

[61] R. Youatt, *Interspecies Politics: Nature, Borders, States* (Ann Arbor: University of Michigan Press, 2020).

interspecies practices, corroborating his argument that IR is a form of interspecies politics.

The interconnected relations between species have long been recognised in Indigenous cosmologies, resulting in the incorporation of the non-human in Indigenous communities' politics and laws.[62] Hence, many argue that it is time for us to mobilise and learn from Indigenous struggles and knowledge how to relate to more-than-human worlds and the Earth itself.[63] Accordingly, there has been an effort to reveal how the non-human can affect global politics and how it can inform the building of a multispecies IR. To do this, many scholars have looked to decentre the human in IR through post-humanist, new materialist, and relational approaches which expose the complex entanglement between humans, other beings, and things from which we depend, and with which and by which we are constituted.[64] For instance, Fishel[65] demonstrates how trees and forests can serve as metaphors of how IR and our societies can be rebuilt. The author describes how forests keep the health of their ecosystems in check and what that might teach us about the importance of reflexivity. She explains how trees benefit from systems of "social security" that not only protect individual trees, but also the forest and ultimately the planet, underlining the importance of solidarity and interdependency. Fishel draws a parallel between the self-destruction of greedy organisms in a forest and the behaviour of humans and highlights the importance of multispecies relationships for the health and survival of the tree (singular and global). Recognising the significance of forests in securing life, but also learning from them, she stresses, has the potential to counter IR's human-centred worldview, to foster political and scientific engagement with non-human life, assure the planet's survival, and thus the survival of world politics.

Meijer[66] explores the global implications of different forms of oppression through a multispecies perspective that highlights the prevalence of anthropocentrism when theorising structural injustice. She argues that analysing structures of injustice through a non-anthropocentric lens is fundamental to understand the oppression of human and non-human animals. The interconnected vulnerability between species and the influence of non-human life in world politics has been repeatedly illustrated in public health crises, including the COVID-19 pandemic.[67] In this regard, Cudworth

---

[62] Pereira and Gebara, "Where the Material and the Symbolic Intertwine".

[63] Inoue, "Worlding the Study of Global Environmental Politics".

[64] Burke and Fishel, "Across Species and Borders"; E. Cudworth and S. Hobden, "Posthuman International Relations: Complexity, Ecology and Global Politics", in *International Relations in the Anthropocene*, ed. Chandler, Müller, and Rothe, 233–249.

[65] S. Fishel, "The Global Tree: Forests and the Possibility of a Multispecies IR", *Review of International Studies* 49, no. 2 (2023): 223–240.

[66] E. Meijer, "Global Injustice and Animals: Towards a Multispecies Social Connection Model", *International Relations* 37, no. 3 (2023): 497–513.

[67] A. Burke, "Interspecies Cosmopolitanism: Non-Human Power and the Grounds of World Order in the Anthropocene", *Review of International Studies* 49, no. 2 (2023): 201–222; E. Cudworth and S. Hobden, "Animalising International Relations", *International Relations* 37, no. 3 (2023): 398–422.

and Hobden[68] describe how the pandemic underlines humans' animality, calling for an "animalisation" of IR and proposing several ways through which the incorporation of non-human animals in the discipline, both conceptually and in its institutions and practice, can be achieved. Beyond that, Burke[69] has advanced the foundations for an interspecies cosmopolitanism that stresses the need to acknowledge the agency, power, and moral value of both human and non-human lives, as well as the relations between them and the processes which constitute them. The aim is to overcome the limitations of cosmopolitanism's humanism which, while important, is largely inadequate for the Anthropocene. In doing so, Burke[70] searches for the needed balance between the recognition that no being is strictly autonomous and the importance of the principle of autonomy to "defend individuals against all kinds of violations". He contends that these principles can bolster political projects and institutions that ensure the rights of both ecosystems and individuals, while addressing the planetary crisis, which the current political, institutional, and economic landscape is doing "imperfectly and unjustly".[71]

Scholars are ultimately suggesting a renewed approach to IR that recognises individuals as embedded in overlapping larger structures and living systems. This approach tries to creatively transcend "the possessive individualism and Enlightenment humanism" in political theory[72] and leads scholars to envision different modes of agency, representation, democracy, and justice,[73] as well as new foundations for a discipline which has developed around the supposed exceptionalism of "an animal which does not think it is an animal",[74] and thus struggled to achieve justice for all Earth beings.[75]

Following this overview of how the Anthropocene has been framed as a significant topic within IR and related fields, the next section will demonstrate how and why a greater scholarly engagement with the Amazon region would allow for advancing the existing, still largely theoretical literature on IR and the Anthropocene.

---

68 Cudworth and Hobden, "Animalising International Relations".

69 Burke, "Interspecies Cosmopolitanism".

70 Burke, "Interspecies Cosmopolitanism", 215.

71 Burke, "Interspecies Cosmopolitanism", 203.

72 D. Celermajer et al., "A Political Theory for a Multispecies, Climate-Challenged World: 2050", *Political Theory* 51, no. 1 (2023): 41.

73 A. Schapper and C. Bliss, "Transforming Our World? Strengthening Animal Rights and Animal Welfare at the United Nations", *International Relations* 37, no. 3 (2023): 514–537; Burke and Fishel, "Across Species and Borders"; Cudworth and Hobden, "Animalising International Relations"; T. Fougner, "Animals and Diplomacy: On the Prospect for Interspecies Diplomacy", *International Relations* 37, no. 3 (2023): 449–474.

74 Cudworth and Hobden, "Animalising International Relations", 414.

75 Celermajer et al., "A Political Theory for a Multispecies, Climate-Challenged World".

A Research Agenda on IR and the Amazon in the Anthropocene

The Amazon region is home to the planet's largest contiguous tropical forest and river system, storing approximately half of all tropical forest carbon while harbouring 10 per cent of the Earth's known species. The rivers and streams of the Amazon are home to the largest number of freshwater fish species on Earth, and the region's forests possibly host one-quarter of all land plants and animals. Amazonian ecosystems provide livelihoods for the region's Indigenous, riverside, and urban populations. They also sustain traditional local cultures and are critical components of the life-support system that makes the world habitable, providing services such as water cycling and climate regulation. The region is also home to hundreds of ethnic groups, hosting an extraordinary diversity of social and cultural life.[76] However, there is mounting scientific evidence that the biome is losing its resilience and may be approaching a tipping point, whereby the rainforest will evolve to savannah in large parts of the region.[77] This should be critical for IR, a discipline explicitly concerned with security and survival, as the dieback of the Amazon could have irreversible implications not only on local populations and regional ecological processes, but also on the stability of the Earth system as a whole.[78] Amazonian ecologies are powerful agents that, if irrepressibly disrupted, can rapidly and abruptly perturb the international system.

Despite that, few researchers in the field have engaged with the Amazon. The scant attention to the region in IR reveals significant power dynamics related to the limited interest in environmental issues[79] and a critical neglect of the role played by the Global South in the making of international relations.[80] Yet it is necessary to explore the synergies that can arise from centring the Amazon in ongoing discussions about IR's shift towards the Anthropocene. In the following sections, we show how the region brings to the fore issues that, up until now, have remained absent from post-anthropocentric IR, but can advance its agenda in important ways, making the discipline better equipped to understand this new proposed geological epoch and tackle the analytical, normative, and political challenges that it poses. Simply, we ask how can in-depth investigations into the Amazon contribute to advance post-anthropocentric approaches in IR.

---

[76] SPA, *Amazon Assessment Report 2021*.

[77] B. M. Flores et al., "Critical Transitions in the Amazon Forest System", *Nature* 626, no. 7999 (2024): 555–564.

[78] SPA, *Amazon Assessment Report 2021*.

[79] J. C. Pereira, "The Limitations of IR Theory regarding the Environment: Lessons from the Anthropocene", *Revista Brasileira de Política Internacional* 60, no. 1 (2017): e0187.

[80] Picq, "Rethinking IR from the Amazon".

Relational, Socio-Cultural Approaches to Land Governance and the Bioeconomy

A significant topic that merits research by IR scholars concerns the collective and multicultural decision-making experiences taking place in the region. These are archetypal examples of a multicultural democracy in action, i.e., spaces wherein public policies have been collectively designed and implemented from the bottom-up, seeking to resist managerialist and exploitative initiatives being imposed from the top-down. Holding a strong transformative potential, these initiatives emerged through the construction of horizontal governance schemes that bring together representatives of various Indigenous associations and federations, and members of peasant communities, NGOs, and governmental agencies. They provide us with a privileged entry point to illuminate and explore the practices and possibilities of post-anthropocentric politics. For instance, in the north-western region of the Amazon, Indigenous groups often govern their shared territories collectively, guided by ecological principles that subordinate political and social practices to the cycles of nature. In doing so, they exemplify an instance wherein political decisions and territorial management strategies are constantly negotiated with the non-human entities that inhabit the region,[81] rooted in the concept of "forestzenship"—the idea that "the forest and its peoples constitute a polis, a more-than-human community endowed with proper political subjectivity", with each playing a role in the preservation and advancement of the community as a whole.[82] Studying these experiences can provide valuable insights for promoting socio-cultural and ecosystem connectivity within and beyond the Amazon and, concomitantly, exploring the possibilities of scaling-up these projects when thinking about more-than-human cosmopolitan ethics of governance and democracy for the Anthropocene. Within the region, eco-socio-cultural connectivity is understood as "the structural and functional connection amongst natural areas which are important for the maintenance of the integrity of ecosystems, local communities, life supporting systems and regional economies".[83]

[81] M. von Hildebrand and V. Brackelaire, *Guardianes de la selva: Gobernabilidad y autonomía en la Amazonia colombiana* [Guardians of the forest: Governability and autonomy in the Colombian Amazon] (Bogotá: Gaia Amazonas, 2012); M. I. Valderrama González, M. I., *Principales retos para la gobernanza de los Territorios Indígenas del Noroeste Amazónico* [The main governance challenges in the Indigenous Territories of the Northwest Amazon] (Bogotá: Gaia Amazonas, 2019).

[82] S. Barca, *Forces of Reproduction: Notes for a Counter-Hegemonic Anthropocene* (Cambridge: Cambridge University Press, 2020), 24.

[83] ANA, *Contributions to the 1st Draft of the New Post 2020 Global Biodiversity Framework (GBF) from the North Amazon Alliance* (2022), accessed May 6, 2024.

Furthermore, the ecological corridors of the Amazon[84] provide relevant examples of alternatives to prevailing perspectives and practices, particularly when considering transformative governance and Anthropocene security. These examples encourage a reconsideration of both governance and security within spatial and temporal dimensions, intra- and interspecies relationships, as well as across the borders of diverse knowledge systems. For example, the regional project to create the Andes–Amazon–Atlantic Corridor brings together multiple rights, stake, and knowledge holders to establish the world's largest ecological corridor. The project reframes discussions around Amazon protection, shifting the focus from national sovereignty to collective responsibility. Furthermore, it proposes an alternative governance framework founded on the acknowledgement of the intricate interdependence between human well-being, non-human living beings, and ecosystem resilience.[85]

Similarly, experiences within Amazonian socio-bioeconomies could yield valuable insights for reconceptualising world politics beyond modernist, conventional economic-growth-driven capitalist, and techno-managerial approaches to human/non-human relations, towards more transformative practices rooted in feminist, decolonial, and care perspectives that enhance ecological and ontological security. Socio-bioeconomies focus on transformative values, namely democracy, economic justice, welfare, and the recognition of diverse ways of knowing, reconciling development and ecosystem conservation. The bioeconomy is often seen as a way to achieve global sustainability. Yet it is generally associated with ideas like ecological modernisation and technological development, or plantation economies, organised around commodity production for economic expansion through the destruction of biodiversity via pulping and burning. In contrast, Amazonian socio-bioeconomies rest upon the inseparability of the material and symbolic values of the forest, prioritising life, diversity, multispecies relations, community engagement and empowerment, participatory decision-making processes, and the sustainable management of nature.[86]

---

[84] G. Rioja Ballivián, *Mapiense Número 2* (2015), accessed 6 May, 2024. https://www.herencia.org.bo/webdocs/publicaciones/Mapiense-2.pdf; M. A. Morales Mite, "Human – Wildlife interactions in the Sangay – Podocarpus Connectivity Corridor, Andes of Southern Ecuador: A socio-ecological and geographic characterization" (master's field practicum report, University of Florida, 2020); Pereira and Terrenas, "Towards a Transformative Governance of the Amazon"; Ministerio de Ambiente y Desarrollo Sostenible – Parques Nacionales Naturales de Colombia, Ministerio del Ambiente, and Ministerio del Ambiente – Servicio Nacional de Areas Naturales Protegidas por el Estado, *El programa trinacional de conservación y desarrollo sostenible del Corredor de Áreas Naturales Protegidas [The tri-national conservation and sustainable development program of the Protected Natural Zones Corridor] PNN* (no date), accessed 6 May, 2024. https://d2ouvy59p0dg6k.cloudfront.net/downloads/historias_de_frontera_1.pdf; Wildlife Conservation Society, *Integrated Management Project of the Putumayo-Içá River Basin. Stakeholder Engagement Plan* (2021), accessed 6 May, 2024. https://documents1.worldbank.org/curated/en/159001622049620200/pdf/Stakeholder-Engagement-Plan-SEP-Integrated-watershed-management-of-the-Putumayo-I%C3%A7%C3%A1-river-basin-P172893.pdf.

[85] Pereira and Terrenas, "Towards a Transformative Governance of the Amazon".

[86] J. C. Pereira et al., "Baniwa Chilli Pepper: Indigenous Spirituality, Socio-Bioeconomy, and Sustainability", in *Handbook of Traditional Spirituality and Sustainability*, ed. J. M. Luetz, W. Leal Filho, and P. D. Nunn (Springer, forthcoming).

The production and commercialisation of the Baniwa chilli pepper of the Alto Rio Negro Indigenous Territory in the Amazon exemplifies the transformative potential of regional socio-bioeconomies. These are community-driven, focusing on the population's priorities in marketing. Moreover, production is carefully managed to prevent disrupting community activities, with plans to expand production inclusively. Diversity is emphasised, with peppers being vibrant and varied, contrasting with monoculture. For Baniwa women—who play a significant role in the project—caring for the pepper plants is akin to caring for family, symbolising protection for the entire community. The production and sale of these peppers exemplify Indigenous leadership in land management, challenging colonial structures. Also, the transition from local consumption to commercialisation reflects intercultural collaboration, merging Indigenous and non-Indigenous knowledge.[87] These regional experiences prompt us to focus on the "forces of reproduction", i.e., the subjects and agents that sustain humanity by nurturing the biophysical environment essential for life. This challenges the Western-centric forces of production, which historically have relied on the destruction of more-than-human worlds.[88] As such, they offer an alternative to existing economic models and power structures. Indeed, researchers and practitioners of world politics can learn numerous lessons from socio-bioeconomies, harnessing encounters between diverse worlds to develop responses to the planetary ecological crisis.

More-Than-Human Personhood and Community Formation

Learning from Amazonian Indigenous relations with non-human communities can further enlighten our discussions on sovereignty and understand what it entails to expand this concept beyond the state and human relations. New forms of relating to animals and integrating them into legal systems arise from Indigenous treaty relations and practices of sovereignty. Indigenous communities in North America are evolving the right to self-determination through their holistic conceptions of sovereignty and the establishment of treaty relations with one another and with settler governments.[89] Traditionally, Indigenous groups in North America have also established treaties between humans and non-human animals to support co-existence and ensure mutual well-being.[90] Similar dynamics to the ones described in terms of treaty making by North American Indigenous groups are also present in the practices of Amazonian Indigenous groups. The Kichwa, for instance, have very similar ways of relating to

[87] Pereira et al., "Baniwa Chilli Pepper".

[88] Barca, *Forces of Reproduction.*

[89] S. Lightfoot and D. MacDonald, "Treaty Relations Between Indigenous Peoples: Advancing Global Understandings of Self-Determination", *New Diversities* 19, no. 2 (2017): 25–39.

[90] A. Mitchell, "Revitalizing Laws, (Re)-Making Treaties, Dismantling Violence: Indigenous Resurgence Against 'the Sixth Mass Extinction'", *Social & Cultural Geography* 21, no. 7 (2020): 909–924; Lightfoot and MacDonald, "Treaty Relations Between Indigenous Peoples".

other beings to the treaty relationships outlined by Simpson,[91] characterised by bonds rooted in empathy and care.[92] Engaging with these relationship dynamics, which stem from a relational building of the self, could inspire alternative approaches to IR's conceptions of political treaties whose main goal is not to attribute possessions or rights, or which stem from a traditional conception of sovereignty.

Furthermore, engaging with experiences of personhood and people-making stemming from more-than-human relations in the Amazon could help IR scholars further the discussion and theories on more-than-human security, specifically on answering the question of who counts as a person (worthy of security), and thus challenge the anthropocentric bases of IR and security studies. It also provides lessons on the understanding of non-human animals as conscious beings with motivations and agency who experience insecurity and what that entails politically. While, in his article, Leep[93] explores notions of spiritual personhood in Ojibwe practices and obligations in safeguarding wolves, many examples of cosmovisions which take the consciousness and subjectivity of animals, and which are worthy of exploration, arise from Indigenous communities in the Amazon. Communities such as the Yanomami, the Tukano, the Xerente, the Maxineru, the Uitoto, and the Achuar see humans as interdependent and interconnected and believe that animals, like humans, have souls and spirits.[94] Like these, many other Amazonian groups not only attribute subjectivity to animals (and things), but also human traits and feelings.[95] This attribution of subjectivity to the non-human aligns with Amerindian theories of personhood, according to which persons are "beings that possess a soul", which "allows for awareness of oneself and of others". In Kichwa thought, for instance, non-human animals have the same level of complexity and consciousness as human animals. This contrasts with the Western view that humans have a "uniquely complex 'inner life' or consciousness".[96] For them, and for many other Amerindian communities, subjectivity stems from the relations in which a being (human or not) exists and which are possible due

[91] L. B. Simpson, *Dancing on Our Turtle's Back: Stories of Nishnaabeg Re-creation, Resurgence, and a New Emergence* (Kolkata: Adivaani, 2012).

[92] J. Reddekop, "Against Ontological Capture: Drawing Lessons from Amazonian Kichwa Relationality", *Review of International Studies* 48, no. 5 (2021): 857–874.

[93] Leep, "Introduction to the Special Issue".

[94] A. M. Simpson (Mitchell), "Healing the Planet: Traditional Spiritual Beliefs and Sustainable Management of Ecosystems in the Amazon Forest, Colombia", *Practical Theology* 15, no. 5 (2022): 432–444; Inoue, "Worlding the Study of Global Environmental Politics"; P. Descola, "Estrutura ou Sentimento: A Relação com o Animal na Amazônia" [Structure or feeling: The relationship with the animal in the Amazon], *Mana* 4, no. 1 (1998): 23–45.

[95] D. Kopenawa and B. Albert, *The Falling Sky: Words of a Yanomami Shaman* (Cambridge: The Belknap Press of Harvard University Press, 2013); E. Viveiros de Castro, *A Inconstância da Alma Selvagem e Outros Ensaios de Antropologia [The wild soul's inconstancy and other anthropological essays]* (São Paulo: Cosac & Naify, 2002); F. Santos-Granero, *The Occult Life of Things: Native Amazonian Theories of Materiality and Personhood* (Tucson: University of Arizona Press, 2009).

[96] Reddekop, "Against Ontological Capture", 863.

to each being having a soul. These assumptions then inform their relations with the non-human.[97]

Hydrological Connectivity and the Transnational Politics of Freshwater Ecosystems

Another important topic pertains to the protection of the waters and welfare of Amazonian riverside populations. Whenever focusing on water-related issues, IR has privileged either an engagement with its geopolitical dimension and potential role as a threat to national security and regional instability or, alternatively, adopted a human-oriented and rights-centred perspective stressing the need to secure the right to water at the global level. The Amazon demands going beyond that and engaging the politics of freshwater ecosystems and hydrological connectivity. Rivers are politically contested entities permeated by more-than-human power relations and cultural-political hierarchies; they are living entities and territories formed, inhabited, and dynamically altered by assemblages of beings.[98] Yet, obscured by the majesty of the forest and the threat of its depletion, protecting the region's aquatic ecosystems has been historically neglected in favour of terrestrial conservation. This resulted in a lack of data and management structures to protect them.[99]

Indigenous relational knowledge and practices on water use and conservation can provide valuable insights into how to respond to the challenge of water security in the Anthropocene. Indigenous peoples around the world have been contesting freshwater management models based on Western ontologies and epistemologies.[100] A deeper engagement with Amazonian water governance could also bring into the fore the security and well-being of Amazonian riverside populations, who are often neglected by decision-makers and international assistance schemes, frequently focused exclusively on Indigenous communities. Interestingly, some artisanal fishing communities within the region engage in daily communicative interactions with other species, as they often interpret the sounds and behaviours of fish, or rely on them to forecast weather patterns. By acknowledging the personhood of fish, these communities have

---

[97] E. Kohn, "How Dogs Dream: Amazonian Natures and the Politics of Transspecies Engagement", *American Ethnologist* 34, no. 1 (2007): 3–24; Reddekop, "Against Ontological Capture".

[98] C. Houart, J. Hoogesteger, and R. Boelens, "Power and Politics Across Species Boundaries: Towards Multispecies Justice in Riverine Hydrosocial Territories", *Environmental Politics* 34, no. 1 (2025): 49–69.

[99] M. Macedo and L. Castello, *State of the Amazon: Freshwater Connectivity and Ecosystem Health* (Brasília: WWF Living Amazon Initiative, 2015).

[100] M. Parsons and K. Fisher, "Indigenous Peoples and Transformations in Freshwater Governance and Management", *Current Opinion in Environmental Sustainability* 44 (2020): 124–139.

also established ethical criteria for fishing practices. However, the voices of these local communities have been excluded from decision-making processes.[101]

The Politics and Interspecific Implications of Environmental Militarisation

Finally, engaging with the Amazon can help IR better appreciate the interspecific character and implications of environmental militarisation. Over the past few years, there has been a clear tendency to associate environment-related issues with the security agenda, as issues such as climate change, climate-induced migration, or natural catastrophes became increasingly salient in national and international security strategies.[102] Meanwhile, it is also important for IR to explore the other side of this coin and engage with the implications of environmental militarisation. As global military expenditures continue to rise, the environmental destruction provoked by the military sector, especially due to its extensive energy usage, has become increasingly apparent. Other devastating environmental consequences result not only from conflict itself but also military training, and include, among others, biodiversity degradation, the destruction of cultural resources, or even the widespread contamination of ecosystems.[103] Despite that, the Amazonian region has been militarised through processes that are often framed as something necessary and justified by either the usage of its borders by transnational organised crime, the need to involve the military in conservation projects, or the territorial disputes between Guyana and Venezuela. Apart from the potentially devastating damages that militarisation can provoke on human and non-human communities in the region, this process also disturbs the delicate balance between other actors involved in governing the region, from local communities to non-governmental organisations.

This is a crucial but underexplored theme for those interested in advancing the agenda of IR in the Anthropocene. Understanding the more-than-human vulnerabilities and insecurities arising from the presence of the armed forces in the forest in recent years,[104] and locating the strategies that resist them, call for in-depth

---

[101] R. Boelens et al., "River Lives, River Movements. Fisher Communities Mobilizing Local and Official Rules in Defense of the Magdalena River", *The Journal of Legal Pluralism and Unofficial Law* 53, no. 3: 458–476.

[102] For instance, the United States National Security Strategy released in 2022 explicitly considers that "[o]f all of the shared problems we face, climate change is the greatest and potentially [most] [*sic*] existential for all nations", while NATO's 2022 Strategic Concept frames climate change as "a defining challenge of our time, with a profound impact on Allied security".

[103] A. K. Jorgenson, B. Clark, and J. E. Givens, "The Environmental Impacts of Militarization in Comparative Perspective: An Overlooked Relationship", *Nature and Culture* 7, no. 3 (2012): 314–337.

[104] J. C. Pereira and E. Viola, "From Protagonist to Laggard, from Pariah to Phoenix: Emergence, Decline, and Re-emergence of Brazilian Climate Change Policy, 2003–2023", *Latin American Policy* 15, no. 3 (2024): 400–422; J. Corredor-Garcia and F. López Vega, "The Logic of "War

studies that shed light on the regional dynamics of military territorial management, its embeddedness in broader political, economic, and social contexts, and how it might fundamentally reshape conservation policy, and public policy more generally, in the longer run. Addressing this necessitates moving the discipline beyond environment-related securitisation and pushing it to explore the effects of militarisation outside exceptional moments, and the potential roles that the military can play when it comes to advancing post-anthropocentric forms of governance that frame socio-cultural connectivity, rather than sovereignty, as the primary referent of security. While the logistical capacity of the military places them in a privileged position to address the humanitarian consequences of natural catastrophes in the short-term, Amazonian countries have also displayed a recurrent interest in militarising the region. Unpacking the logics and assumptions that underpin these attempts and the consequences resulting from the presence of military troops in the region is particularly important in this context, especially given the link already established between militarism, biodiversity degradation, and climate change.

Conclusion: Closing the Gap Between IR and the Amazon

The Anthropocene puts forward a major challenge for IR. By destabilising its foundational assumptions and categories, it pushes the field to move beyond its disciplinary and empirical boundaries in search for alternative ways of knowing and doing that are more attuned to this new geological epoch. In this chapter, we have argued that the Amazon is a pivotal arena for advancing post-anthropocentric perspectives on world politics. After reviewing the IR literature on the Anthropocene, we illustrated the analytical, ontological, and ethico-political insights that can result from a more systematic and reflexive engagement with the region. We focused on IR's encounter with regional relational, socio-cultural approaches to land management and the bioeconomy, local understandings of personhood and community formation transcending the human, the dynamics of freshwater ecosystems and hydrological connections, and the multifaceted character and consequences of environmental militarisation. Apart from enriching its theoretical and conceptual apparatus, we posit that bringing the Amazon into IR is also important to facilitate the transition of the broader research agenda on the Anthropocene from theory to practice.

While Amazonian perspectives and experiences can contribute to advancing existing research within IR on the planetary ecological crisis, ongoing debates in the field can also enrich our comprehension of the challenges and opportunities associated with conserving the Amazon. For instance, studies on wildlife crime in

---

on Deforestation": A Military Response to Climate Change in the Colombian Amazon", *Alternatives: Global, Local, Political* 49, no. 4 (2023): 325–343; Pereira and Viola, *Climate Change and Biodiversity Governance in the Amazon.*

the Amazon that examine animal cultures, akin to the research conducted by Lappe-Osthege and Duffy[105] on bird migrations and related crimes along the Adriatic and eastern Mediterranean Black Sea Flyway, have the potential to encourage the development of more robust responses to criminal activities within the region. Meijer's[106] multispecies social connection model can aid in mapping the global structures of more-than-human injustices, which connect populations in the Amazon with distant others. This will offer a clearer perspective on human and animal oppression that transcends state boundaries, shedding light on the varied responsibilities of different kinds of actors in promoting change and guiding their efforts towards goals that can help protect the rainforest and its inhabitants. If incorporated into Amazonian policy, Dryzek and Pickering's[107] notion of "ecological reflexivity" could catalyse the development of adaptable management frameworks capable of real-time adjustments driven by insights and feedback from Indigenous and other local communities, scientists, and other stakeholders. These frameworks would monitor and asses the outcomes of diverse interventions, adapting strategies as needed. Such an approach holds significant relevance, especially given the emerging signs of savannisation in regions of the forest in a climate-changed world.[108]

In addition to that, exploring how IR's theoretical frameworks can shed light on Amazonian politics would also be important. For instance, securitisation theory[109] could be particularly helpful to help us understand how regional security has been discursively constructed as an existential threat and how certain aspects, such as the proliferation of transborder organised crime, have been framed as a challenge requiring exceptional measures that facilitate a militarisation of the region. This theory could also be mobilised to examine how certain governments and actors sought to de-securitise the need to protect Indigenous cultures, forests, or endangered species and thus justify the rollback of environmental policies.[110] Differently, the idea of human security and its focus on both direct and structural vulnerabilities faced by the most vulnerable communities, including its link to environmental degradation or climate change, but also stressing food, health, or economic security,[111] could be an entry point to investigate the everyday livelihoods and multidimensional insecurities experienced by riverside and Indigenous communities. Feminist, post-structural, and postcolonial research strategies would be particularly helpful to understand the gendered and racialised dynamics of environmental governance,

---

[105] Lappe-Osthege and Duffy, "International Relations and the Non-Human".

[106] Meijer, "Global Injustice and Animals".

[107] Dryzek and Pickering, *The Politics of the Anthropocene.*

[108] L. P. Sales, M. Galetti, and M. M. Pires, "Climate and Land-Use Change Will Lead to a Faunal "Savannization" on Tropical Rainforests", *Global Change Biology* 26, no. 12 (2020): 7036–7044.

[109] B. Buzan, O. Wæver, and J. de Wilde, *Security: A New Framework for Analysis* (Boulder: Lynne Rienner Publishers, 1998).

[110] L. Hansen, "Reconstructing Desecuritisation: The Normative-Political in the Copenhagen School and Directions for How to Apply It", *Review of International Studies* 38, no. 3 (2012): 525–546.

[111] UNDP, *Human Development Report 1994: New Dimensions of Human Security* (New York, 1994).

unpack how certain modes of environmental knowledge help normalise and perpetuate top-down, command-and-control, anthropocentric modes of thinking, knowing, and relating to Amazonian ecologies and non-human beings, as well as explore the resistance strategies put forward by local actors. Along a similar line, critical constructivist modes of inquiry focused on narratives could be used to analyse how different actors advance contending views of security and development and explore the extent to which these promote practices that are attuned to the planetary real.[112] Ultimately, closing the gap and advancing this research agenda requires exploring the synergies between IR and the Amazon against the backdrop of the Anthropocene.

[112] M. McDonald, "Discourses of Climate Security", *Political Geography* 33 (2013): 42–51.

Chapter 3
Diplomacy in the Anthropocene: The Climate and Cyber as Threat Multipliers

Ana Flávia Barros-Platiau, Carlos Henrique Rubens Tomé Silva, Guilherme Lopes da Cunha, Fábio Henrique Granja e Barros, Paulo Eduardo Saraiva Câmara, Jorge Gomes do Cravo Barros, and Laura Cristina Feindt Urrejola

Introduction

Climate change and the failure to address its deleterious impacts can be considered an existential threat to humanity. Likewise, we can no longer live detached from cyberspace, which affects our societies and all aspects of our relations, including those with whom we have no direct relations. In this sense, life has become truly

A. F. Barros-Platiau (✉) · C. H. R. T. Silva · P. E. S. Câmara · L. C. F. Urrejola
University of Brasilia (UnB), Brasília, Brazil
e-mail: anabarros@unb.br

G. L. da Cunha
Brazilian War College (ESG), Rio de Janeiro, Brazil
e-mail: guilhermelopes11@icloud.com

F. H. G. Barros
Federal Accounting Court (TCU), Brasilia, Brazil
e-mail: fabiohg@tcu.gov.br

P. E. S. Câmara
Brazilian Defence College (ESD), Brasilia, Brazil

J. G. do Cravo Barros
Brasilia, Brazil

L. C. F. Urrejola
Universidad Nacional de La Plata, Buenos Aires, Argentina

M. Robson (ed.), *The Anthropocene, Ontopolitics, and International Relations*,
https://doi.org/10.1007/978-981-96-2925-1_3

"global"[1] because what happens in cyberspace can also affect anyone, anywhere, at any time in real life. In a nutshell, land, sea, air, space, and cyber realms are connected in an unprecedented manner. Thus, daily life and cyber activities have become deeply interwoven. As a consequence, the line between the natural and the digital worlds becomes blurred.[2]

In this context, the multiple connections between the climate and cyber agendas range from integrated risk assessment to energy transition, encompassing issues of food, water, and health security, as well as human rights. These connections have myriad implications, but not all of them are positive.

On the one hand, technological tools allow users to have access to big data from all over the world, analyze them in real-time, update existing global climate models, and improve climate-related decision-making processes. Frequently, information and knowledge about global climate change are quickly available in the digital world. In this sense, multilateral negotiations about mitigation of greenhouse gas emissions (GHG), adaptation and funding can benefit from the best available information.

On the other hand, the same tools can be used by malicious actors for warfare and terrorism, among other security threats. For instance, they can be used to destabilize a government, disrupt critical infrastructure, and promote violence or panic, thus challenging statecraft and multilateralism. In sum, the technologies used in cyberspace can be extremely helpful or dangerous when they are related to climate change agendas.

Additionally, the climate and cyber agendas share common features, since they are deeply intertwined,[3] overlapping, polycentric, and complex (Box 1). These features emphasize the need for integrated approaches to address them in more flexible and agile manners. In this vein, when the United Nations Secretary-General, António Guterres, took office, he acknowledged that climate change and disruptive technologies, especially artificial intelligence (AI), were two huge regulatory agendas that needed to be developed within the United Nations framework.

---

[1] E. Kavalski (2015, p. 6) refers to "global life" as a concept in line with I. J. Davidson-Hunt and F. Berkes (2003, p. 54) as a "human-in-ecosystem" perspective on the study and practice of IR which recognizes the "mutual influence of ecological and social processes, instead of treating social and ecological as linked but separated domains". Consequently, "The notion of global life therefore elicits that all human interactions are embedded in and made possible by complex global interconnections" (2015, p. 5). When he mentions non-human, he refers to the around (the environment), while our text refers to robots and the artificial world, including artificial intelligence. E. Kavalsky (2015) 'Introduction: Inside/Outside and Around: Observing the Complexity of Global Life,' in E. Kavalski (Ed.) World Politics at the Edge of Chaos: Reflections on the Complexity of Global Life, Sunny Press, pp. 1–20; I. J. Davidson-Hunt & F. Berkes (2003) Nature and society through the lens of resilience: toward a human-in-ecosystem perspective. Navigating social-ecological systems: Building resilience for complexity and change, 53, 82.

[2] M. Carr & F. Lesniewska (2020) 'Internet of Things, cybersecurity and governing wicked problems: learning from climate change governance,' *International Relations,* 34(3), pp. 391–412.

[3] R. Setola; E. Luiijf & M. Theocharidu (2017) 'Critical Infrastructures, Protection and Resilience,' in R. Setola et al. (Eds) *Managing the Complexity of Critical Infrastructures*, Springer, pp 1–18; G. Manrique; B. Dupont & C. Shearing (2022) 'Resilient infrastructures in the Anthropocene,' CrimRxiv, 32 p., available at: https://www.crimrxiv.com/pub/v7695ksu/release/1, accessed on 4 December, 2024.

Box 1—Points in Common for Climate Change and Cyberspace Challenges

1. They can amplify other threats
2. They do not respect sovereign rights
3. They are limitless. Their risks have been underestimated so far
4. They are caused by human action
5. System interactions are not governed by rules
6. States alone do not control them
7. They are shaping our future and the Planet
8. They correspond to tipping points and cascading effects
9. They can be used with geopolitical interests
10. They affect human security and human rights on an unprecedented scale
11. Their evolution is outpacing the multilateral institutional framework
12. Their impacts are unpredictable[4]
13. They need to be governed on a planetary scale
14. They constitute enormous threats to human security
15. To address them, we need exponential diplomatic changes

Although there is a plethora of recent reports on how scientific and technological innovations shape our lives, the International Relations scholarship connecting the challenges of the digital world to the Anthropocene paradigm is still incipient. While the Anthropocene experts focus more on the natural/sociopolitical world and climate change,[5] technology experts tend to focus more on human–machine interactions, AI, and cyberspace.

Because they disregard national boundaries and sovereignty rights, climate change and cyberspace reinforce some key features of the present global governance challenges in the Anthropocene epoch: pervasive influence, global interconnectedness, integration of realms, and reciprocal influence. So, this chapter aims to connect these two extremely complex and politicized agendas by claiming that diplomacy and geopolitics cannot be disconnected from climate and cyber threats combined in polycentric arenas. Therefore, one of the main arguments put forward is that the recent initiatives from the United Nations, various agencies, states, and different actors from the private sector are still largely insufficient for global governance.

---

[4] Climate and cyber security face the same challenge of "autonomy". While climate depends on the forces of nature, for cyber see: N. S. Fouad (2022) 'The non-anthropocentric informational agents: Codes, software, and the logic of emergence in cybersecurity,' *Review of International Studies*, 48(4), pp. 766–785.

[5] See for example: the debate on the enormous risks and threats due to the overstepping of the "planetary boundaries". J. Röckstrom et al. (2009) 'A Safe Operating Space for Humanity,' *Nature*, 461, pp. 472–475, available at: https://www.nature.com/articles/461472a, accessed on 4 December, 2024; J. Röckstrom (2024) 'The Tipping Points of Climate Change — and Where We Stand,' *TED Talk*, 15 August 2024, available at: https://www.youtube.com/watch?v=Vl6VhCAeEfQ, accessed on 4 December, 2024; A. Boncheva, E. Delgado (2025) 'Anthropocene, Climate Change, and Sustainability,' in A. Boncheva & E. Delgado (Eds.) *Transition to a Safe Anthropocene in the Asia–Pacific*. Springer.

While cyberspace gains around 320 million terabytes daily, the governance of the Internet remains concentrated in the United States, not without objection and countermeasures.[6] Moreover, some big tech companies[7] and individual billionaires, nowadays named "tech executives" can be more powerful than sovereign states. This is truly a multistakeholder chessboard[8] as the United Nations and the 2024 Declaration from the Global Partnership on AI (GPAI)[9] recognized. Consequently, we can define the current international order by the term of "three trillions", which corresponds to NVidia's stock market value[10]; a little more than the estimated amount of US dollars of military expenditure in 2023[11]; and more than the Brazilian total GDP (the eighth biggest economy in the world).[12]

In this context, authorities, geopoliticians, and diplomats must adapt to dystopias that are yet to emerge. Therefore, updating diplomacy concerns how decision-makers must be prepared to harness the opportunities to address insecurity from climate change and cyberspace simultaneously. Considering the "climate change emergency",[13] "climate failure"[14] and its impacts, as well as the expansion of cyberspace, we pose the following question: in the context of a new great powers' competition, how do we address climate and cyberdiplomacy in line with the Anthropocene paradigm?

---

[6] Some countries have strategies for reducing their dependency on the US governance model, such as Russia and China.

[7] V. Mendes (2022) 'Big Tech Firms and Climate Change. Mapping the Low-Carbon Vested Interests of Alphabet, Amazon, Apple, Meta,' PhD diss., IR Institute, University of Brasilia.

[8] M. Carr & F. Lesniewska (2020), ibid.

[9] GPAI has a partnership with the OECD. The 2024 GPAI New Delhi Declaration has the goal "to continue engaging in constructive dialogue through the integrated partnership to achieve our shared vision of a future where AI is used for the good of all, leaving no one behind and promoting peace, prosperity, and sustainable development". GPAI (2024), '2024 GPAI New Delhi Declaration', *Global Partnership on Artificial Intelligence*, available at: https://gpai.ai/gpai-new-delhi-declaration-2024.pdf, accessed 23 November, 2024.

[10] Reuters (2024) 'Nvidia's stock market value hits $3 trillion for the first time,' Reuters, 5 June, 2024, available at: https://www.reuters.com/technology/nvidias-stock-market-value-hits-3-trillion-first-time-2024-06-05/, accessed on 23 November, 2024.

[11] N. Tian et al. (2024) 'Trends in World Military Expenditure, 2023,' *Stockholm International Peace Research Institute*, SIPRI Fact Sheet, available at: https://www.sipri.org/publications/2024/sipri-fact-sheets/trends-world-military-expenditure-2023, accessed on 4 December, 2024.

[12] According to Forbes India, the GDP in 2024 was worth US$ 2.33 trillion. See https://www.forbesindia.com/article/explainers/top-10-largest-economies-in-the-world/86159/1, accessed on 23 November, 2024.

[13] J. Verbitsky (2023) 'Antarctica and Problems of Antarctica Governance in an Era of Climate Emergency,' in A. F. Barros-Platiau; F. Queiroz & P. Câmara (Eds.) *Estudos Antárticos: Dimensão Humana, Ciência e Geopolítica*, Coleção Antártica, O Sistema do Tratado da Antártica, v. III, Brasilia: FUNAG, available at: https://funag.gov.br/biblioteca-nova/produto/1-1252, accessed 4 December, 2024.

[14] Diplomatic failure to address the climate change threats has been widely debated by UN authorities and academia. The most recent example is the CoP 29, for which the final deal did not secure adequate funding for the security and safety of the majority of humanity in most developing countries.

To answer this question, the chapter starts with the Anthropocene paradigm, bringing up challenges for global governance from an International Relations perspective. Certainly, "grand challenges" for global governance abound: regulating the malicious uses of cyberspace; managing the climate system; preventing the outburst of pandemics; and regulating the biotechnological rapid evolution,[15] among others. Then the chapter highlights the current context of new great powers' competition to claim that geopolitical priorities hamper diplomatic cooperation efforts to tackle climate and cyber threats. Therefore, the second part acknowledges that the solutions implemented are still largely inadequate for the problems the scientific community can predict.[16] If climate change can be considered a "threat multiplier",[17] the same is true for cyberspace. In this way, in both spaces, digital and natural/social, accelerated and uncontrolled developments jeopardize human security, whether in the realm of individual and political freedoms or the conditions necessary for the maintenance of life on Earth. Finally, although the importance of the private sector is underlined, the chapter focuses on state actors and diplomacy. This choice goes in line with Prantl's and Goh's question: "how do international actors mobilize power and exercise statecraft within a world that gives them less control over the outcomes they want to achieve?".[18]

Part 1—The Anthropocene Paradigm

The Anthropocene paradigm highlights the transformation of human activity into a dominant force driving environmental changes and planetary dynamics. The causes of the Anthropocene are directly linked to human actions, whose scale has significantly increased and accelerated since the Industrial Revolution. Human activities have reached a magnitude capable of altering the dynamics of various processes essential to maintaining life on the planet,[19] such as the rapid melting of the cryosphere and climate anomalies, such as the "2024 super El Niño". Although

---

[15] O. Young (2021) *Grand Challenges of Planetary Governance. Global Order in Turbulent Times*. Cheltenham: Edward Elgar.

[16] IPCC (2023) 'Summary for Policymakers,' in *Climate Change 2023: Synthesis Report*, IPCC: Geneva, pp. 1–34; WEF (2024) 'Global Risks Report 2024', *World Economic Forum*, available at: https://www.weforum.org/publications/global-risks-report-2024/, accessed on 4 December, 2024; R. S. Litwak (2023). 'Tripolar Instability: Nuclear Competition Among the United States, Russia, and China,' *Wilson Center Publisher*, available at: https://www.wilsoncenter.org/book/tripolar-instability-nuclear-competition-among-united-states-russia-and-china, accessed on 4 December, 2024.

[17] S. Goodman & P. Baudu (2023) 'Climate Change as a 'threat multiplier': History, uses and future of the concept,' *Center for Climate and Security*, Briefer no. 38, available at: https://climateandsecurity.org/2023/01/briefer-climate-change-as-a-threat-multiplier-history-uses-and-future-of-the-concept/, accessed on 4 December, 2024.

[18] J. Prantl & E. Goh (2024) 'Cultivating 'Strategic Diplomacy' in the Indo-Asia-Pacific and Beyond', *Strategic Diplomacy homepage*, available at: https://www.strategicdiplomacy.net/, accessed on 23 September, 2024.

[19] J. Röckstrom (2024), ibid.

not formally recognized in the field of Geosciences,[20] the explanatory value of this concept has been widely acknowledged in the social sciences and humanities for emphasizing the anthropogenic deleterious impacts on the Earth system.[21] Therefore, this paradigm raises fundamental questions about how societies manage natural resources, how they treat vulnerable populations, and how they define economic development in the context of global risks.

The concept of Anthropocene equally corresponds to an epoch of accumulation of scientific knowledge and resources, potentialized by artificial intelligence, machine learning, and other technological innovations that transform our problems and solutions, thus reshaping our capabilities to deal with the climate change agenda.[22] Borders that seemed insurmountable such as the deep seabed, the Amazon,[23] and the Arctic[24] become more economically viable with technological innovations to exploit them[25] and growing demand for critical minerals. Antarctica may be next.

One telling example is the current demand for batteries to accelerate the decarbonization of the transportation sector.[26] This demand also engenders environmental damages that exacerbate competition among several stakeholders, producing a "fortress mentality" and a "pernicious spiral of harm".[27] Therefore, geography and demography are continuously transformed by the "scramble for what is left",[28] and by climate change threats,[29] but with new tools that only a few stakeholders can afford to pay for, due to growing concentration of power in their hands.

However, the most emphasized aspect of humanity's profound impact on the planet is climate change. Some observers are so worried about the tipping points that they

---

[20] A. Witze (2024) 'Geologists reject the Anthropocene as Earth's new epoch—after 15 years of debate,' *Nature*, 6 March, 2024, available at: https://www.nature.com/articles/d41586-024-00675-8, accessed on 13 October, 2024.

[21] D. Simangan (2024) 'Peace in the Anthropocene,' *Oxford Research Encyclopedia of International Studies*, available at: https://x.gd/AqMRd, accessed on 4 December, 2024.

[22] J. Prantl et al. (2024) Building Capabilities for Earth System Governance, *Elements Earth System Governance*, pp. 35–48, Cambridge: Cambridge University Press.

[23] F. Queiroz; G. Cunha & A. F. Barros-Platiau (2023) (Eds.). *Brazil in the Geopolitics of Amazonia and Antarctica.* New York: Lexington.

[24] For the time being, Antarctica is a sovereign-free space in which economic activities are forbidden under the Antarctic Treaty System, except for tourism. Norway, Russia and the United States are the big players interested in oil drilling and seabed mining in the Arctic.

[25] See, for instance, 'Motion for a resolution on Norway's recent decision to advance seabed mining in the Arctic', European Parliament, 31 January, 2024, available at: https://www.europarl.europa.eu/doceo/document/B-9-2024-0095_EN.html, accessed on 23 November, 2024.

[26] A. Klimašauskaitė et al. (2024) 'Could deep-sea mining sustain sustainability? The international seabed authority and the UN 2030 Agenda,' *Resources Policy*, 98, 105332, available at: https://doi.org/10.1016/j.resourpol.2024.105332, accessed on 4 December, 2024.

[27] R. White (2014) 'Environmental Insecurity and Fortress Mentality,' *International Affairs*, 90(4), pp. 835–851.

[28] M. Klare (2012) *The Race for What's Left: The Global Scramble for the World's Last Resources*, Metropolitan Books.

[29] M. Klare (2019) *All Hell Breaking Loose: The Pentagon's Perspective on Climate Change*, Metropolitan Books.

adopted the concept of "existential threat". The emission of greenhouse gases has altered—and will continue to alter in the coming decades—the Earth's atmospheric composition. The increased heat retention in the atmosphere is causing a rise in average temperatures, significant changes in the planet's climate dynamics, ocean acidification, sea-level rise, and other effects that may prove catastrophic within this century.[30]

Other effects of the Anthropocene include biodiversity loss, with the potential for mass species extinction due to habitat destruction and over-exploitation of resources, alterations in the carbon, nitrogen, and phosphorus cycles, which are essential for ecosystem functioning, air and ocean pollution (especially from plastic and microplastic accumulation), and the release of toxic and radioactive waste into the environment. Moreover, new bacteria, fungi, and viruses, for example, may cause diseases yet unheard of.

In order to connect all these points, one of the most powerful models to address these transformations was proposed by Johan Rockstrom et al. (2009) as "planetary boundaries".[31] Certainly, human actions have impacts that extend far beyond the environmental agenda. Ecosystem degradation and the increased frequency and severity of extreme weather events, when combined, induce human, social, economic, and political impacts. From a human perspective, these include increased respiratory diseases due to air pollution, and vector-borne diseases. In the social sphere, they include food insecurity, population displacement (climate migrations), and rising inequality. Economically, the Anthropocene perturbations negatively affect sectors such as agriculture, fisheries, and tourism, as well as leading to soaring adaptation and mitigation costs as time goes by. Politically, they exacerbate conflicts over natural resources, both domestically and internationally. These are just a few examples of how the Anthropocene—and climate change, in particular—are disrupting the planetary equilibrium experienced over the past 10,000 years during the Holocene. However, all these alerts are not new. According to the United Nations, between 2005 and 2015, some 700 thousand people died, 23 billion were displaced and 1,5 billion were affected due to natural disasters.[32] Consequently, in a scenario of global governance "business as usual", global temperature may rise beyond 1,5 C in a few decades and the 2030 UN Agenda sustainable development goals will most likely not be attained.

However, the Anthropocene also corresponds to the increase in human capabilities to use natural and artificial resources,[33] thereby creating a complex nexus between

[30] IPCC (2023), ibid.

[31] J. Röckstrom et al. (2009), ibid.

[32] UN (2024) 'Nova iniciativa da ONU procura reduzir riscos de desastres naturais através da IA' [New UN initiative seeks to reduce natural disaster risks through AI], *ONU News*, 24 August, 2024, available at: https://news.un.org/pt/story/2024/08/1836551, accessed on 21 September, 2024.

[33] J. Prantl et al. (2024) Building Capabilities for Earth System Governance, *Elements Earth System Governance*, Cambridge: Cambridge University Press.

environmental pressures and political tensions.[34] Put bluntly, a small part of humanity has been aggravating damages that cannot be fixed easily, but also creating new risks and threats that remain largely unregulated.[35]

Additionally, the negative effects of the Anthropocene are not evenly distributed around the world. Historically, states have prioritized national economic growth, often at the expense of environmental sustainability. Industrial development based on the burning of fossil fuels, combined with the legacy of colonial exploitation, has created inequalities in both responsibility and the capacity to address sustainability problems. More recently, the proliferation of globalized production and consumption chains have intensified the use of natural resources and worsened greenhouse gas emissions. In this context, high-income nations of the Global North are historically responsible for the majority of environmental degradation, while the associated impacts disproportionately fall on countries in the Global South, thus raising questions of intra- and inter-generational responsibilities. In this simplistic framework, the roles of China, India, and Brazil are not clear. As a consequence, the concept of "humanity" needs to be revisited to highlight who causes the problems, in terms of responsibility and liability, and who the victims are.

Although the concept of Anthropocene corresponds to urgency and irreversibility, responsibility, and material complexity,[36] gaps in international institutions' ability to address these issues reveal the difficulty of coordinating effective responses on a global scale. The unfolding of the climate regime after 29 conferences of parties is one telling example, so the Anthropocene epoch is rather characterized by the failure to cooperate. Furthermore, the Anthropocene affects both global stability and the legitimacy of governance institutions. Hence, it introduces a new paradigm for International Relations approaches, requiring all agents to rethink how they address complex challenges that transcend borders and disciplines. The current scholarship concerning the Anthropocene is rather focused on the human responsibility towards the global changes that have been accelerated since the end of Second World War.[37] Also, in high-income countries and other big economies, rapid economic growth has worsened the scientific alerts that humanity is not prepared for the forthcoming climate tipping points and their cascading effects.

---

[34] K. Moorthy (2023) *Climate Change and Geopolitics: The Impact on Global Security: A Complex Nexus: How Environmental Pressures and Political Tensions Converge.* Kindle eBook. ASIN: B0C37X7VYR.

[35] G. Neaher et al. (2024) 'Responsible and Safe AI: A Primer for Policymakers in the Global South,' *Global Task Force for Security and Development, Predictive Analysis*, New America, and Igarapé Institute, available at: https://igarape.org.br/en/responsible-and-safe-ai-a-primer-for-policymakers-in-the-global-south/, accessed on 4 December, 2024.

[36] F. Zelli & P. Pattberg (2016) 'Conclusions: Complexity, Responsibility and Urgency in the Anthropocene,' in F. Zelli & P. Pattberg (2016) *Environmental Politics and Governance in the Anthropocene*, Routledge, pp. 231–242.

[37] J. McNeill & P. Engelke (2016) *The Great Acceleration: An Environmental History of the Anthropocene since 1945.* Harvard University Press; R. Blasiak (2020) 'Blue Acceleration: Our dash for ocean resources mirrors what we've already done to the land,' *The Conversation*, 24 January, 2020, available at: https://theconversation.com/blue-acceleration-our-dash-for-ocean-resources-mirrors-what-weve-already-done-to-the-land-130264, accessed on 4 December, 2024.

Put bluntly, technological innovations have allowed human activities to disrupt planetary dynamics, thereby exacerbating monopolies, inequalities, and injustice on a global scale. Considering the velocity of the development, deployment, and improvement of AI for example, in addition to the fact that this technology can amplify the biases of its designers[38] and bring risks from the digital to the real world, the Anthropocene is also an alert for human–machine interaction. Therefore, human security, positive peace, and ecological justice are better understood through the critical analysis of the limitations of the current multilateralism architecture and the concept of Anthropocene.[39] So, "smart and green policies are essential".[40]

Additionally, the prevalence of economic interests over social and environmental rights not only persisted since the creation of the United Nations but has been consolidated with time. This process results not in "prosperity for all" but in unfair and unsustainable growth on a global scale. As a result, multilateral efforts have failed in many realms, such as curbing greenhouse gas emissions and implementing adaptation measures. As Dalby (2020)[41] puts it, the failure to quit the model of globalization based on fossil fuels corresponds to new threats that do not derive from past rivalries. Consequently, faced with the many factors and dimensions involved, assessing the past in an attempt to predict the future through extrapolation is an insufficient strategy.[42] In this context, what is the connection of diplomacy and geopolitics to the Anthropocene, climate change, and cybersecurity challenges? We argue that there is a double answer: one concerning power politics that hinder international cooperation (part 1) and threats related to climate change and cybersecurity (part 2).

The New Great Powers' Competition

International peace and security are both the heart of geopolitics and diplomacy. They are also the reason why the United Nations was created in 1945. After the end of the Second World War, a bipolar order was established opposing the United States to the Soviet Union. Ensuring security in this context required sophisticated negotiations that led to the Cold War and a technological race to produce nuclear bombs and reach the moon, for example. Although there were dangerous moments due to their competition, the "Western liberal order" prevailed for decades and offered stability to

---

[38] G. Neaher et al. (2024), ibid.

[39] D. Simangan (2024), ibid.

[40] S. Himanshu & S. Kreft (2024) 'Artificial Intelligence – Help or Harm for the Climate?' *United Nations University*, 28 September, 2024, available at: https://unu.edu/ehs/series/artificial-intelligence-help-or-harm-climate, accessed on 4 December, 2024.

[41] S. Dalby (2020) *Anthropocene Geopolitics: Globalization, Security, Sustainability*. University of Ottawa Press.

[42] S. Abranches (2017) *A era do imprevisto: a grande transição do século XXI [The Age of the Unforeseen: The Great Transition of the 21st Century]*. Companhia das Letras.

the international system.[43] More recently, the geopolitical landscape became highly dangerous again, due to the 2022 Russian-Ukrainian war, the 2023 Israeli-Palestinian conflict that became a threat of war in the Middle East, and the consolidation of the Sino-Russian axis.

In addition, the rapid evolution of technological innovations reinforced competition, rivalries, and disputes, mainly among the high-tech great powers, corporations, and designers. Among other geopolitical considerations, it convenes to highlight the US-China relations present a content that encompasses: i) the "chip war", which involves both the production chain for semiconductors, and normative platforms, such as the Chips and Sciences Act,[44] paving the way for a tech run; and ii) the Strait of Taiwan disputes, since Taipei manages AI servers' producers.

Moreover, competition for resources prevails, since high-tech products demand raw materials from peripheral countries. Unfair trade and technological advances have damaged peoples' lives in developing and least developed countries, as demonstrated through the Amazonian Beetle *Lamprocyphus augustus* that provides photonic crystal for producing cutting-edge optical computers.[45] Another case is the access of Coltan, a mineral found in Latin America and Africa and used for electronics. This mining activity has resulted in war, conflict and violence,[46] proving that there are no limits to the scramble for raw materials. In the future, the deep seabed, the Arctic and Antarctica will most likely be explored by mining companies and countries as well, following the geopolitical trend of ruthless struggle for natural resources.

The Challenges for a Better Global Governance Framework

To build a better global governance architecture, diplomacy is one of the most relevant concepts and the first challenge. Diplomacy has evolved in myriad different ways, thus "entering uncharted political territories such as climate negotiations or, lately, cyber issues".[47] But while "climate diplomacy" and climate law have evolved

[43] J. Prantl (2013) 'Taming Hegemony: Informal Institutions and the Challenge to Western Liberal Order,' *Chinese Journal of International Politics*, 7(4), pp. 449–482.

[44] The White House (2022) 'CHIPS and Science Act Will Lower Costs, Create Jobs, Strengthen Supply Chains, and Counter China,' *Fact Sheet*, 9 August, 2022, available at: https://www.whitehouse.gov/briefing-room/statements-releases/2022/08/09/fact-sheet-chips-and-science-act-will-lower-costs-create-jobs-strengthen-supply-chains-and-counter-china/, accessed on 23 November, 2024.

[45] M. Ragaei; A.-K. Sabry & A. Abdel-Rahman (2016) 'Insect's Photonic Crystals and Their Implications,' *Bioscience Research*, 13(1), pp. 15–20; M. M. Amin (2015) 'A Amazônia na geopolítica mundial dos recursos estratégicos do século XXI [The Amazon in the global geopolitics of strategic resources in the twenty-first century],' *Revista Crítica de Ciências Sociais*, 107.

[46] R. Sierra (2020) 'On the Origins of the State: Stationary Bandits and Taxation in Eastern Congo,' *Journal of Political Economy*, 128(1), pp. 32–94.

[47] A. Barrinha & T. Renard (2017) 'Cyber-diplomacy: The making of an international society in the digital age,' *Global Affairs*, 3(4–5), pp. 353–364.

significantly since the 1992 Earth Summit, this is not true for the cyberdiplomacy case. However, defining cyberdiplomacy is not an easy task. The literature review has shown a significant gap since most papers do not distinguish cyber from traditional diplomacy.[48] The case of cyberdiplomacy is particularly interesting because diplomats were not necessarily trained to work with cutting-edge technologies, notably in developing countries. Insofar, cyberdiplomacy encompasses not only public affairs, but also geopolitical and strategic affairs, due to their effects in military, intelligence, and business realms.

What is more, nowadays diplomats have to work with "private authorities",[49] that is, stakeholders that have increasing influence in formal and informal decision-making processes, ranging from big techs to individual terrorists. Their motivations vary from economic gains to political influence. On top of that, the private sector has a bigger role to play in Anthropocene geopolitics, because it concentrates more financial and technological resources than most sovereign states. For private agents, the digital transformation of companies demands more adaptability, resilience, and innovation for sustainable and effective competitive advantages linked to a "culture of sustainability".[50] In addition, technological innovations also increase opportunities for the proliferation of startup companies that can be informal and harder to identify. So, engaging different and asymmetrical non-state stakeholders in the making of a more equitable global governance is mandatory but also an enormous uphill struggle. Finally, as Harari (2024) and Pavie (2018)[51] put it, innovators in general are not liable for their products. Compared to carmakers or pharmaceutical companies, some states or billionaires are still free to exploit outer space and the moon, the deep seabed and cyberspace. In sum, a first enormous challenge for decision-makers is the architecture and agency for the global governance of cyberspace.

A second challenge regards existing and emerging threats, because it is no longer possible to identify who the opponents are, nor when and how they will attack. Of course, traditional rivalries remain, and strategic alliances can be formed and shaped according to the geopolitical priorities of key players. In this new scenario, the predators are hard to identify,[52] be they state or non-state actors, while the prey can be anyone, including people in the most powerful countries, or even adolescents.

---

[48] A. Attatfa; K. Renaud & S. De Paoli (2020) 'Cyber Diplomacy: A Systematic Literature Review,' *Procedia Computer Science*, 176, pp. 60–69.

[49] J. Green (2014) *Rethinking private authority. Agents and entrepreneurs in global environmental governance.* Princeton University Press.

[50] M. Minevich (2023) *Our Planet Powered by AI: How We Use Artificial Intelligence to Create a Sustainable Future for Humanity.* Wiley.

[51] Y. Harari (2024) *Nexus: Uma breve história das redes de informação da Idade da Pedra à inteligência artificial* [A brief history of information networks from the Stone Age to artificial intelligence], Companhia das Letras; X. Pavie (2018) *L'innovation à l'épreuve de la philosophie* [Innovation put to the test of philosophy], Paris: PUF.

[52] A. Cohen (2024) *The Cyber Predators. Dark Personality and Online Misconduct and Crime*, Cambridge University Press.

In this context, the UN Secretary-General defends that the UN has a key and legitimate role to play regarding AI governance. Therefore, the Global Initiative on Resilience to Natural Hazards through AI Solutions gathered various UN agencies (the International Telecommunication Union (ITU), the UN Environment Programme (UNEP), the UN Framework Convention on Climate Change (UNFCCC), the Universal Postal Union (UPU), and the World Meteorological Organization (WMO).[53]

Additionally, confidence-building measures (CBMs) can be considered a third challenge, because they are fundamental in cyberspace, which can be compared to a new territory to be explored. Although there are some CBMs for ICTs, they are largely insufficient, as is the case for AI. So, this place is where diplomatic skills are most necessary in the multiple agenda-setting processes. While the GGE/UNODA recommended eleven voluntary, non-binding norms of responsible State behavior concerning ICTs, most of them can be replicated to AI related to climate change and to the peace and security agendas.

In this vein, in March 2024, the UN General Assembly (UNGA) adopted a landmark Resolution on "Seizing the opportunities of safe, secure and trustworthy artificial intelligence systems for sustainable development".[54] The draft text was proposed by the United States and co-sponsored by 123 states, with the UNGA's acknowledgment that AI has the potential to benefit the 17 Sustainable Development Goals from the 2030 Agenda. According to the UN Secretary-General António Guterres, the governance of AI needs a multistakeholders' forum for dialogue and a scientific body, comparable to the International Panel on Climate Change (IPCC) at the climate change regime, so that information and knowledge can be shared.[55] Consequently, in September 2024, the UNGA adopted the Pact for the Future (Resolution A/RES/79/1),[56] with chapters on: Sustainable development and financing for development; International peace and security; Science, technology and innovation, and digital cooperation; Youth and future generations; and Transforming global governance.[57]

---

[53] See 'Global Initiative on Resilience to Natural Hazards through AI Solutions' homepage, available at: https://x.gd/RxXtG, accessed on 23 November, 2024. The Focus Group on Artificial Intelligence for Natural Disaster Management (FG-AI4NDM) collaborates also with the AI for Early Warnings for All (EW4All) sub-group.

[54] United Nations General Assembly 78th Session, Resolution A/RES/78/265, 'Seizing the opportunities of safe, secure and trustworthy artificial intelligence systems for sustainable development,' available at: https://daccess-ods.un.org/access.nsf/Get?OpenAgent&DS=A/RES/78/265&Lang=E, accessed 21 September, 2024.

[55] GZERO Media (2024) 'Why António Guterres believes the UN should lead on AI: Exclusive interview', *Youtube video*, available at: https://www.youtube.com/watch?v=_52Lic8GIzQ, accessed on 21 September, 2024.

[56] United Nations General Assembly 79th Session, Resolution A/RES/79/1, 'The Pact for the Future', available at: https://daccess-ods.un.org/access.nsf/Get?OpenAgent&DS=A/RES/79/1&Lang=E, accessed on 26 September, 2024.

[57] UN 'Summit of the Future' homepage, available at: https://x.gd/iMyzN, accessed on 26 September, 2024.

A fourth challenge is the need of security policies and diplomacy to adapt to "global futures",[58] in the sense of the complexity and interconnectedness of the fate of our civilization concerning climate change and the agenda for peace.[59] Although sustainability, innovation, and cooperation are taken as the fundamental solution to peace and security on a global scale, they lag far behind in the "real world" current diplomatic negotiations. More effective and comprehensive strategies are the key to enhance global security, stability, agility, and sustainability. Put bluntly, multilateral talks need to encompass the engagement of multiple stakeholders, conflict and disasters prevention, climate-sensitive frameworks, and confidence-building efforts. However, one of the higher hurdles is still geopolitics, since the most powerful actors are currently more engaged in competition and war, not in cooperation for the vital interest of humankind. This is particularly worrisome when climate change and cyber activities may be considered threat multipliers.

Part 2—Climate Change and Cyber Activities as Threat Multipliers

The Climate Change Failure

Climate change exacerbates existing risks and threats, particularly in terms of global security and stability, by worsening social, economic, and political problems, thus amplifying their consequences, and making them more difficult to manage.[60] Rising temperatures, changing precipitation patterns, and extreme weather events, reduce the availability of essential resources like water and food. Sea level rise is another grave issue to be addressed urgently, in a context of deglaciating land and deicing seas. As a result, there may be more extreme weather events and climate anomalies in the coming years.

This can escalate internal and regional disputes over resources, intensifying preexisting conflicts. Desertification and the degradation of agricultural lands make productive land more valuable, increasing competition among communities and countries, which can fuel social and political tensions. Volatility in the prices of food and other basic resources can lead to economic crises and public dissatisfaction, creating scenarios of protests, unrest, and potential conflicts.[61]

The increased frequency and intensity of extreme weather events put millions of people at risk, creating greater demand for humanitarian aid and paving the way for

---

[58] G. Neaher et al. (2024), ibid.

[59] D. Simangan (2024), ibid.

[60] R. Bartnik (2021) 'Diagnosis of threats to the European Union's internal security,' *Humanities and Social Sciences*, 28(3), pp. 7–18.

[61] V. Koubi (2019) 'Climate change and conflict,' *Annual Review of Political Science*, 22(1), pp. 343–360.

the proliferation of humanitarian crises that may be difficult to mitigate.[62] From a societal perspective, it is urgent to forge a legal concept for those people who are forced to flee.[63] However, some people cannot leave their countries, who are most often invisible to multilateral policymakers. There are also the animals that need to adapt or flee, and those that cannot move far away from their natural habitat. Particularly vulnerable regions, such as small islands and low-lying coastal areas, will most likely exacerbate regional and global migration flows.

Developing countries, which have less capacity for adaptation and recovery, are the most affected by the effects of climate change. Frequent climate disasters hinder economic development, disrupt supply chains, and increase poverty, potentially leading to political and social instability.[64] Climate change also amplifies threats to human health, such as the spread of vector-borne diseases (malaria, dengue) and problems related to air quality and the availability of drinking water. Pressure on health systems can engender humanitarian crises in already vulnerable countries.

As an international security threat, climate change can exacerbate vulnerabilities in sensitive geopolitical areas. For instance, growing competition for control of strategic resources, such as the Arctic, where melting ice has opened new shipping routes and access to oil and gas reserves, can heighten tensions among major technological powers. Furthermore, changes in oceanic and atmospheric circulation patterns can also affect energy security, as they alter the availability of natural resources.[65] The threat-multiplying effect of climate change is particularly relevant in the field of International Relations, as it demands more effective global cooperation to mitigate the impacts, both through climate adaptation and mitigation policies, as well as through security and international governance strategies that are more integrated and adaptable to the context of a warming planet.[66]

---

[62] S. Lynn et al. (2010) 'Introducing a 'Hot System'-approach to tipping points in humanitarian crises,' in X. Shen; T. E. Downing & M. Hamza (Eds.) *Tipping points in humanitarian crises: from hot spots to hot systems*, Studies of the University: Research, Counsel, Education', United Nations University, UNU-EHS no. 13/2010, pp. 14–21, available at: https://collections.unu.edu/view/UNU:1880#viewAttachments, accessed on 4 December, 2024.

[63] UNHCR (2022) *People Forced to Flee. History, Change and Challenge.* United Nations High Commissioner for Refugees, available at: https://x.gd/YImYh, accessed on 4 December, 2024; F. Gemenne et al. (2022) 'Forced displacement related to the impacts of climate change and disasters,' Reference Paper for the 70th Anniversary of the 1951 Refugee Convention, *UNHCR*, available at: https://x.gd/YImYh, accessed on 4 December, 2024.

[64] L. Signé & A. Mbaye (2022) 'Renewing Global Climate Change Action for Fragile and Developing Countries,' *Global Economy & Development at Brookings, Working Paper*, 179; R. McCarney & J. Kent (2020) 'Forced displacement and climate change: Time for global governance,' *International Journal*, 75(4), pp. 652–661.

[65] J. Markowitz (2020) *Perils of plenty: Arctic resource competition and the return of the great game*. Oxford: Oxford University Press.

[66] C. Deere-Birkbeck (2009) 'Global governance in the context of climate change: the challenges of increasingly complex risk parameters,' *International Affairs*, 85(6), pp. 1173–1194; M. Robin (2021) *La fabrique des pandémies. Préserver la biodiversité, un impératif pour la santé planétaire* [The making of pandemics. Preserving biodiversity, an imperative for planetary health*]*, Paris: La Découverte.

Cyberspace as a Threat Multiplier

Cyberspace expands quickly as more people are connected to services of all sorts. Nowadays, around 5,3 billion people have access to the Internet, which means that they are also more exposed to threats lurking on the web. However, "Our digital world is one of divides", since the asymmetries related to "region, gender, income, language, and age groups" remain.[67] For instance, while some 89% of people in Europe are connected, only 21% of women in low-income countries have access to the Internet.[68] In this scenario of digital divides, the 2024 GPAI New Delhi Declaration commits to "global multistakeholder cooperation" and "human-centric AI" through an inclusive approach concerning the process of dialogue for standard-setting.

Undoubtedly, technological innovations are powerful tools to manage information, planning, and human choices related to the sustainable use of resources. Also, big data and AI allow to update climate models with unprecedented accuracy. However, as mentioned in the previous section, climate crises can also be exacerbated by cyber vulnerabilities, because they intensify existing risks and make structural problems more difficult to manage.[69] In this context, the 2024 GPAI Declaration "acknowledges the emerging risks and challenges posed by AI systems, particularly advanced AI systems, including those related to safety and security and potential malicious uses; mis- and disinformation; harmful biases leading to discrimination; lack of transparency and fairness; lack of level playing field; risks to intellectual property and personal data protection; threats to human rights and the wellbeing of children; risks for environmental sustainability and democratic values; widening digital divides between and within countries; and transforming the future of work".[70] In brief, there are myriad ways in which technology affects the processes of socialization, thus escaping from the control of state actors.

Cyber activities were first considered as a threat to security in a traditional manner. Then, other agendas such as the climate case were included in the multilateral debate, with the support of Secretary-General Guterres. In a geopolitical mode, the use of information is part of the competition between great powers, thus shaping the future of wars. Currently, AI is being used in unmanned crafts, drones, killer robots, and others, for example, in Air Combat at DARPA's AlphaDogfight Trials.[71] So, cyberspace is now part of the battleground. Because it has no clear boundaries and claimants, cyberspace is so hard to define. If the case of the cyberattack in Estonia in 2007 is

---

[67] EOSG (2023) 'A Global Digital Compact — an Open, Free and Secure Digital Future for All,' Our Common Agenda Policy Brief 5, *United Nations Executive Office of the Secretary-General (EOSG)*, available at: https://www.un-ilibrary.org/content/papers/10.18356/27082245-28, accessed on 4 December, 2024.

[68] ITU (2022) 'Measuring Digital Development: Facts and Figure 2022,' International Telecommunication Union.

[69] R. Bartnik (2021), ibid.

[70] GPAI (2024), ibid; N. S. Fouad (2022), ibid.

[71] A. Pope; J. Ide & D. Mićović (2023) 'Hierarchical Reinforcement Learning for Air Combat at DARPA's AlphaDogfight Trials,' *IEEE Transactions on Artificial Intelligence* 4(6), pp. 1371–1385.

considered a tipping point,[72] recent cyberattacks have shown the quantity and quality of potential threats. One telling example was the Stuxnet worm put into the Iranian nuclear facilities in 2010.

Moreover, cybersecurity involves complex issues that go beyond the simple protection of networks and systems, encompassing national security, digital sovereignty, state behavior, the protection of critical infrastructure, and international cooperation to address transnational threats. Competition for control over data and digital resources in a context of growing interdependence and globalization can raise issues related to an ongoing "digital arms race". Through cyber espionage, states and organizations seek to obtain strategic information, such as government data or industrial secrets. Countries are increasingly investing in offensive and defensive cyber capabilities, thus increasing the risk of escalation in cyber conflicts. The governance of Internet infrastructure, and emerging technologies, such as artificial intelligence (AI), 5G and 6G, generates disputes among major powers, particularly between the United States, China, Russia, and the European Union, over global standards and norms.

Effective prevention of malicious users in the field of cybersecurity requires global normative standards, confidence, and technological innovation. Therefore, great powers have mixed interests in cooperating for the new governance of artificial intelligence systems for sustainable development, although all 193 UN members signed the Resolution A/RES/78/265.[73]

While international law has evolved significantly since the creation of the United Nations, technologies related to cyberspace have evolved much faster. Moreover, the UN Security Council has been malfunctioning since its creation. Therefore, activities that could threaten world security and peace are poorly regulated. For instance, does a cyberattack correspond to an act of war? If yes, what kind of laws should be developed? If non-state actors can threaten public and private institutions, what should be the role of diplomats? How should they engage with the private sector? Should AI and ICTs be considered weapons? If so, how do we redefine the use of force? Likewise, norms, rules, and principles that regulate state activities to prevent the illegal use of force need a new architecture to tackle challenges associated with cyberspace activities. Because cyberspace is a contested political space shaped by values, interests, and norms,[74] it is extremely difficult to regulate activities related to breakthrough technologies.

The transnational and asymmetric nature of cyber threats combined with climate change makes it imperative for states and other stakeholders to develop innovative governance mechanisms capable of addressing the challenges of an increasingly interconnected and digitalized world. Through multilateral cooperation, it is necessary to establish international treaties that regulate behavior in cyberspace, such as the development of international norms for state "cyberconduct", which is currently under discussion at the United Nations.

---

[72] A. Attatfa; K. Renaud & S. De Paoli (2020), ibid.

[73] United Nations General Assembly 78th Session, Resolution A/RES/78/265, ibid.

[74] A. Barrinha & T. Renard (2017), ibid.

Promoting cybersecurity and climate change security, therefore, requires a multidimensional approach that involves the creation of international norms and the promotion of effective multilateral cooperation. Improving resistance to threats associated with climate change and cyberspace implies a new geopolitical framework since the "heartland" of Mackinder (1919) needs a new interpretation. The "cyber" architecture has no heart and no land. Furthermore, the higher the dependency, the higher the risks.[75] Third, power and security are not necessarily assured by military might, since individual hackers can cause big trouble to public and private organizations.[76] So, cyber strategies are necessary to secure national interests[77] and the security of societies.

In the realm of geopolitical tensions, existing conflicts can be amplified by cyber activities, such as cyber espionage, the theft of sensitive data, and attacks on critical infrastructure. Diplomatic and military tensions, when intensified, may trigger cycles of retaliation and escalation. Additionally, states may employ hacker groups or non-state actors to carry out cyberattacks, complicating attribution and making diplomatic responses more difficult. This can escalate conflicts already present in other domains, such as trade or military disputes, by adding a layer of uncertainty and distrust.[78] During times of armed conflict or military tensions, cyberattacks can serve as threat multipliers by disrupting communications, disabling defenses, or creating tactical confusion. This can make local conflicts more destructive or turn a regional war into limitless threats, because cyberspace is connected to critical infrastructures such as power grids, nuclear facilities, telecommunications, transportation, health and financial systems. A cyberattack on any of these networks could magnify existing threats, such as internal political tensions, natural disasters, or economic crises, potentially creating a chain of events that amplifies catastrophes. For instance, in natural disaster scenarios, a cyberattack on emergency response systems, communication networks, or supply chains could worsen the damage and hinder recovery efforts.[79]

As states, businesses, and even military infrastructures become increasingly reliant on digital networks, any cyber vulnerability can multiply the impacts of malicious users. For example, a cyberattack on the healthcare system could worsen a public health crisis, as was observed with ransomware attacks during the COVID-19 pandemic. Another threat concerns the institutional responses to this new technological landscape. Although our "digital future" is expanding rapidly, cyberspace remains poorly understood and regulated. Thus, data monetization enables big tech

[75] UN (2024). *Governing AI for Humanity*, United Nations AI Advisory Body, Final Report, available at: https://www.un.org/sites/un2.un.org/files/governing_ai_for_humanity_final_report_en.pdf, accessed on 4 December, 2024.

[76] A. Cohen (2024), ibid.

[77] A. Attatfa; K. Renaud & S. De Paoli (2020), ibid; A. Barrinha & T. Renard (2017), ibid.

[78] C. Azubuike (2023) 'Cyber Security and International Conflicts: An Analysis of State-Sponsored Cyber Attacks,' *Nnamdi Azikiwe Journal of Political Science*, 8(3), pp. 101–114.

[79] M. Dawson et al. (2021) 'Understanding the challenge of cybersecurity in critical infrastructure sectors,' *Land Forces Academy Review*, 26(1), pp. 69–75.

companies to grow faster than states.[80] One of the consequences is that authoritarian regimes are empowered to control people, thus violating human rights to privacy and freedom of speech, as proposed through the concept of "panspectrum surveillance". For Braman[81]:

> The concept of the panopticon refers to surveillance practices in which the individual subject of surveillance is first identified and then multiple techniques and technologies of observation are directed upon the subject. The use of surveillance techniques as a means of control under modernity is often described as panoptic, and this accurately summarizes the cumulative effect of many of the practices of the bureaucratic state. In the informational state the panopticon has been replaced with the panspectron, in which information is gathered about everything, all the time, and particular subjects become visible only in response to the asking of a question. The panspectron is also a control mechanism, with the additional features that it can manage many more subjects at once, and that the subjects of surveillance never know when, how, or why they might become visible on the panspectral screen.

The evolution from Bentham's panopticon (1995) to a panspectrum approach[82] marked a fast modernization process. It characterized a mechanism for hierarchical control in his first proposition, as proposed originally by Bentham.[83] In modern times, through the Information Communication Technology (ICT), the development of those ideas contributed to give rise to a public traceability and accountability in contemporary social life, where "a collaborative system of shared surveillance, where everyone watches everyone else in favor of morality and good customs" emerges.[84]

Moreover, the growing sophistication of cyberattacks has the potential to negatively impact the global economy, especially in a scenario of market hyper-interconnectivity. Attacks on financial institutions, data breaches, and disruptions to digital supply chains can exacerbate economic crises and increase global inequality, acting as a multiplier of social and political tensions.[85] Moreover, cyber vulnerabilities are aggravated in emerging economies or developing countries, which often

---

[80] S. Zuboff (2019) *The Age of Surveillance Capitalism: The Fight for a Human Future at the New Frontier of Power*. Public Affairs; I. Bremmer (2023) 'A Próxima Superpotência Mundial não é Quem Possam Pensar,' [The Next World Superpower Is Not Who You Think] *TED Talk*, 14 June, 2023, available at: https://www.youtube.com/watch?v=uiUPD-z9DTg, accessed on 4 December, 2024.

[81] S. Braman (2006) *Change of State: Information, Policy, and Power*. The MIT Press.

[82] J. Bentham (1995) 'Bentham's panopticon: Letter I,' in M. Bozovic (Ed.) The Panopticon Writings, Verso, pp. 33–34; S. Braman (2006), ibid.; A. Roca (2020) 'Sociedad de la Información, Sociedad Digital, Sociedad de Control,' *Inguruak – Revista Vasca de Sociología y Ciencia Política*, 68, pp. 50–77.

[83] In the Letter I, written in Crecheff, Russia, in 1787, J. Bentham unveiled his observations on how to optimize hierarchical control through "a space not too large to be covered or commanded by buildings, [where] a number of persons are meant to be kept under inspection." (M. Bozovic, 1995, ibid, p. 33).

[84] R. Cianconi & Y. Lott (2016) 'Do panopticon ao panspectron: uma reflexão sobre as mediações de poder e a materialidade da informação,' [From the panopticon to the panspectron: a reflection on the mediations of power and the materiality of information] *Liinc em Revista*, 12(2), pp. 243–257.

[85] E. Kala (2023) 'Critical Role of Cyber Security in Global Economy,' *Open Journal of Safety Science and Technology*, 13(4), pp. 231–248. O. Gulyás & G. Kiss (2023) 'Impact of cyber-attacks on the financial institutions,' *Procedia Computer Science*, 219, pp. 84–90.

have less capacity to defend against cyberattacks. This multiplies the threats already present in these contexts, such as corruption, political instability, and a lack of cybersecurity infrastructure, further amplifying existing crises.[86]

Extremist groups, such as terrorist organizations, use the Internet for both coordination and the spread of disinformation and recruitment. This amplifies the threat they already pose in the physical world, further complicating counterterrorism efforts. Moreover, extremist groups may use social media to mobilize individuals and incite violence online or physically. Cyber disinformation campaigns are a significant form of threat multiplier, particularly in democracies, where electoral processes and public trust can be undermined. Interference in elections through cyberattacks and the manipulation of social networks to sow social division can exacerbate internal political crises and lead to prolonged unrest.[87]

Cyberspace is also an open space for corruption, trafficking, money laundering, and other illegal activities in the financial/trade sector. These activities are not disconnected from the proliferation of arms dealers, mercenaries, pirates, and militia. Therefore, AI and ICTs can encourage mass investments in unsustainable activities beyond the control of states, such as the case of deforestation in the Amazon region, funded by money from unknown sources. When cyber vulnerabilities are imbricated with other global threats—such as political instability, geopolitical tensions, corruption, terrorism, economic crises, and even natural disasters—they magnify the risks and consequences of these threats. As a result, mitigation, adaptation, and resistance to climate threats also encompass protection from cyber criminality and predatory investments.

Conclusion

Today, life is global if we consider safety and security concerns within national borders, areas beyond national jurisdictions, and cyberspace together. From an International Relations perspective, planetary boundaries and global risks are part of a conceptual framework that corroborate the paradigm of the Anthropocene epoch with scientific empirical evidence drawn from social and political transformations, and big data.

The causes of the Anthropocene epoch lie at the intersection of politics, economics, and the environment, with significant responsibility resting on states and private actors. For this chapter, the Anthropocene paradigm corresponds to a context of scientific alerts about "climate emergencies" combined with the unprecedented production of knowledge and groundbreaking technologies. In this context, global

---

[86] M. Hamidi & B. Singh (2024) 'Designing a Novel Cybersecurity Framework to Prevent Cyber-Attacks with Reference to Least Developing Countries,' *Nanotechnology Perceptions*, 20(S3), pp. 159–165.

[87] N. Komang et al. (2023) 'Cybercrime and Threats to the Electoral System,' *Journal of Digital Law and Policy*, 3(1), pp. 26–37; Y. Pally (2020) 'Cyberattacks and Disinformation Campaigns,' *Journal of Social Political Sciences*, 1(3), pp. 195–206.

sustainability challenges have become central issues for diplomacy and geopolitics. However, the Anthropocene paradigm in International Relations scholarship is frequently silent in terms of connections of the climate agenda to cyberspace. Therefore, the paradigm must be complemented with the diplomatic and geopolitical perspective concerning cyberspace, thus encompassing issues such as security and safety, power projection, the use of force, violence disseminated in societies, and the concentration of resources of power.

Connecting the global climate change debate to the cyberspace one, we contend that diplomacy in the twenty-first century is evolving not only in terms of AI and information technologies, but also in terms of shaping global climate governance. While AI and other technologies contribute enormously to the management of climate change agendas, such as integrated risk assessment, human security, and energy transition, they can also be employed as weapons to destabilize societies and promote violence or panic. Not only do new technologies have clear implications for the future of peace, security, wealth, and health, but also for life on Earth. As a result, although cutting-edge technologies are necessary to accelerate the attainment of the 2030 Agenda's SDGs, they can also bring serious risks and threats.

The potential promises and perils of the poorly regulated use of technologies such as AI and ICTs have been transforming the world society, and thereby the role of authorities in general, and diplomats in particular. Consequently, diplomacy corresponds to innovative transversal activities concerning people, the planet, peace, prosperity, and partnerships, in line with the United Nations framework. However, diplomacy evolves on a multistakeholder chessboard, even though nobody knows who these stakeholders are. Furthermore, due to geopolitical competition, the lack of a multilateral normative framework to regulate behavior in cyberspace creates a dangerous global governance gap. This hinders global cooperation, despite the recent UN efforts under Secretary-General Guterres' lead. In this vein, it makes sense to consider the climate change diplomatic failure with the cyberspace regulatory gaps as threats to international peace and security. Combined, they constitute serious challenges to global governance and our shared future.

Therefore, the main challenge for diplomats is to contribute to the creation of new multistakeholder institutions, inside and outside the UN, and the empowerment of existing ones, so that this current lose-lose spiral in which public authorities are unable to cooperate is replaced. In other terms, diplomacy and geopolitics need to converge to the conviction that cyberspace can be used either to fix or to aggravate the problems associated with the Anthropocene paradigm and climate failure.

This debate connecting climate and cybersecurity has only started, but it is clear that cyberspace gives room for the concentration of wealth and power through technological innovations in the hands of a few, thus deepening current divides worldwide. Who will put responsibility on their shoulders? Who will give voice to the rest of humanity, that is, those who have been disenfranchised so far? Future research is necessary to address all these questions. Looking forward to the desirable global future for all, it is clear that climate and cyber agendas are irreversibly connected.

Chapter 4
Multi-Species Justice and the Anthropocene

Gabriela Kuetting

Many of the most pressing crises facing humankind in the 2020s are connected to the way humans commodify nature and operationalize multi-species relations. Multi-species relations are a fundamental driver of what constitutes the Anthropocene. This chapter aims to provide an overview of how multi-species relations and particularly multi-species justice have been studied in global environmental politics and how the commodified nature of multi-species relations is a core constituent of the Anthropocene. The age of Anthropocene in global environmental politics denotes the historical stage at which ecological processes are largely influenced by human actions and with regards to multi-species relations, the Anthropocene is an era in which non-human species' lives are largely determined by human actions such as, for example, large scale extinctions as a result of habitat destruction. This chapter will trace the historical relations of multi-species relations and then discuss how to arrive at a more multi-species justice in the Anthropocene.

Multi-species relations are particularly visible with regards to industrial farming methods and most strongly present in factory farming. The ecological and carbon footprint of factory farming is immense.[1] The connection between viruses, pandemics and the conditions in which animals are farmed are undeniable and in fact, the majority of viruses develop in factory farms.[2] Thus, regardless of the ethics of multi-species relations, factory farming is a substantial contributor to the global environmental crisis and a massive health hazard. Further, health issues arising from diets based on too much or unsafe meat consumption are rife. Food recalls because of bacterial contamination and resulting deaths are widespread. Yet, the demand for

[1] Food and Water Watch, https://www.foodandwaterwatch.org/wp-content/uploads/2021/03/fs_1806_factory_farms_and_climate_change-web.pdf, 2018, accessed 13 September 2024.

[2] Otte, J.; D Roland-Holst et al., *Industrial Livestock Production and Global Health Risks, PPLPI Research Report Nr 07-09*, June 2007, FAO.

G. Kuetting (✉)
Rutgers University, Newark, NJ, USA
e-mail: kutting@newark.rutgers.edu

M. Robson (ed.), *The Anthropocene, Ontopolitics, and International Relations*,
https://doi.org/10.1007/978-981-96-2925-1_4

cheap meat is insatiable and rising globally as western lifestyles are emulated in other parts of the world with rising incomes. Dairy-free and highly processed vegan products hardly fare any better. Thus, the consequences of factory farming, while bringing cheap food, are at the root of many pressing problems society faces today although this connection is not often made explicit in public debates. In fact, historically, the rise of the fossil fuel economy with ensuing extractivism and intensive agriculture are at the structural root of global environmental degradation and the early stages of the Anthropocene.

While many companion animals are increasingly treated like mini humans with, for example, the rise of cancer or diabetes treatments for pet dogs and cats on a steep increase in western lifestyles,[3] there is no concomitant rise in concern about the welfare of factory farmed animals or animals losing their habitats except by a small number of animal or nature rights activists. Yet, in many ways the commodification of farm animals in factory farms can be described as a colonial form of exploitation of other sentient beings as does the destruction of habitats. This raises the question whether the sentient status of other beings should be relevant as a social/political question or whether the commodification of the nature-society and particularly the human-animal relationship is legitimate. This is of course a normative question as nature-society questions *are* very much commodified. Historically, many groups of social actors have only recently gained representative status or have not gained it yet in a meaningful way—non-white, female, child or Indigenous member of society are a case in point as are non-nationals, LGBTQ or differently abled members. What has brought them more representation or legal equal status (but not necessarily social equal status) is the recognition that they are sentient beings and fellow human beings. The normative question posed here is whether the 'line' of duty of care and representation should be drawn at the 'fellow human being' or 'sentient being' point. This chapter will engage with the socio-political origins of multi-species relations and analyze the basis from which the commodification of other-than-human species is legitimized. It will then engage with the main debate in political theory regarding animal rights, the notion of sentience and whether sentience ought to be considered in nature-society relations. This will be followed by an overview of attempts at assigning rights to nature and whether such an approach is compatible with achieving multi-species justice.

Thus, this chapter is about justice and representation/recognition for sentient beings other than humans, or rather society members endowed with representational rights. Parts of this chapter will focus on animals but does not argue from an animal rights perspective per se as is conventionally understood. Rather, it engages conceptually multi-species relations between sentient beings on the planet beyond human beings—be it in individual species form or as a collective. It also takes an inclusive perspective in that while it is focused on animals, concern with ecosystems is equally included in the analysis. This chapter wants to question what is not usually questioned—the fact that we accept without thinking that rights are only associated with

---

[3] Biospace, www.biospace.com/article/pet-diabetes-care-market-rise-in-prevalence-of-diabetes-in-pet-animals-to-drive-the-market, 2020, accessed 13 September 2024.

represented members of society who can advocate for their rights. Nature itself does not have any recoverable rights. A tree or a fox do not have the capability to engage in acts that would help them enforce their rights (if they had any). In other words, an oak tree or an African elephant cannot engage in deliberative processes, do not have standing in court and of course do not have the skill set to argue their case. This does not mean, however, that they are entirely without rights, but they are unable to participate in the process that accords them or withholds those rights. In the famous words of a famous article, 'trees do not have standing'.[4] As a result of this inability and of the nature-society relations framing the architecture of the western world, a hierarchy of relations exists that ultimately disenfranchises all sentient beings apart from adult humans.

So why is this a topic that has not garnered attention in global politics in general and global environmental politics specifically? It is perhaps not surprising that the rights of sentient beings other than citizens imbued with political and legal rights has not figured prominently in global politics as politics and political science is concerned with 'actors'—and these sentient beings ranging from the disabled, to children and to animals, are by and large not 'acting' in the global system in the traditional sense. We now see underage activists in the Friday for Future movement demanding the right to be heard and to participate, based on the argument that they will one day be citizens with rights and that is in fact their future that is being decided upon by today's actors. In the same way that women, people of color and Indigenous Peoples were invisible in the global system until fairly recently (and continue to be invisibilized), the focus on being able to demand or claim one's rights determines the attention given particular groups. The focus is not on social (or 'natural') groups that cannot vocalize their rights. A polar bear cannot sue for the destruction of its habitat because of a rapidly warming planet. An industrially farmed chicken cannot complain to Health and Safety officers that animal welfare codes have been violated, or, indeed, that it is deprived of its freedom to roam. It cannot protest its slave-like role in the economy either. A pet guinea pig cannot take its owner to court for animal cruelty or for the right to die in dignity for that matter. Political and policy approaches that take the world 'as is' as their epistemological and ontological starting point do not have the capacity to engage with non-human species, or nature or ecosystems for that matter, in the global system as they do not neatly fit the existing analytical categories of these frames.[5] Critical approaches could potentially develop such concepts but attempts to do so are still in their infancy, especially in traditional global environmental governance approaches. In global environmental politics, there is some recognition that there are sentient beings whose existence is severely impacted by decisions that they have no control over. Concerns such as biodiversity, the impact of development on ecosystems, the coloniality of social relations are just a very few examples.

[4] Stone, Christopher. "Should Trees Have Standing? Toward Legal Rights for Natural Objects," 45 *Southern California Law Review* 450 (1972), 450–501.

[5] Cochrane, Alasdair. *Sentienist Politics*. Oxford: Oxford University Press, 2018.

The Evolution of Modern Multi-Species Relations

The roots of contemporary discourses of nature can be traced back to the sixteenth century and the application of the dualist Cartesian lens that defined nature as an object of man's rationality and, in doing so, socially constructed the idea of nature as being an inferior "other" to humans. When referring to nature here, in the context of this chapter and multi-species justice, nature is considered both as an abstract 'ecosystem' and also refers to particular parts of 'nature', i.e. species. This radically new idea had profound implications for those existing ontologies and epistemes that viewed the relationships between 'man', nature and the divine in qualitatively distinct ways from that of modernity. This was seen both within the "Western" world as well as in regions colonized by the Europeans. Given the analytical intention of modern reason to break "the whole" into component parts along the subject-object axis, embedded relationships between man, nature and the divine were disentangled from one another with far-reaching ontological consequences, as Machado Aráoz explains in the following:

> Going back to the Judeo-Christian tradition, the Enlightenment radicalized the separations between the sacred, the human and nature, instituting different ontological regions (the physical and the metaphysical) and establishing a correlative rupture between mind and body and between philosophy, science and religion (Lander, 2000).
>
> Root of all ontological and epistemic fissures, the foundational opposition between Subject and Object radically reconfigures the notion of the 'human' and, with it, that of reality as a whole. The Cartesian Subject crystallizes as the key vector of the entire modern scaffolding, its mechanisms of regulation and social production and its knowledge-power structures thereby radicalizing the separation between Man and Nature. This fundamental split is going to materialize/extend in two logically corresponding directions: the 'de-sacralization' of nature and the 'denaturing' of the human.[6]

The hierarchical dichotomy of human versus non-human found in the Cartesian human–nature dualism manifested itself not only within the European context of the Enlightenment but even more so in the colonization of the rest of the world. It is not only defined the relationship between humans and the biophysical world, but also the relationships between humans. Colonized people were also considered part of the category of "nature", thus demonstrating the expansion of this construct to anthropocentric social relations. While this is a crucial aspect, the confines of this chapter and necessary focus prevent a detailed discussion of this aspect.

The transformation of global nature-society relations can be traced back to 1492 and the colonization of Abya Yala and Turtle Island, territories the Europeans christened the "New World" or the "Americas". With the onset of the "Columbian Exchange", ecological imperialism had widespread effects on colonized landscapes, territories and Indigenous communities, forever altering pre-colonial nature-society relations. Citing the seminal work of Alfred Crosby, Francis explains in the following:

---

[6] Machado Aráoz, Horacio. "La 'Naturaleza' como Objeto Colonial: Una Mirada desde la Condición Eco-Bio-Pólitica del Colonialismo Contemporáneo" ['Nature' as Colonial Object: A View From the Eco-Bio-Political Condition of Contemporary Colonialism] *Boletín Onteaiken* No. 10 (2010), 35–47.

> …Europeans brought pathogens and alien plant and animal species to the New World. With the establishment of empire, they attempted to create 'neo-Europes', taking with them plants and domesticated animals to other regions of the world which were nostalgic remnants of the landscape and ecology they had left behind in Europe. The effects were devastating: unknown diseases such as smallpox-decimated indigenous populations who had no immunity (Crosby 2004: 200). Besides destroying people, this 'portmanteau biota' coupled with the avarice of settler communities steeped in a modern economy that commoditized nature and landscape as private property, also resulted in ecocide. Alien invasive species threatened to wipe out a significant number of indigenous flora and fauna, erasing the 'memory' that the landscape and ecologies had of themselves, in some instances leading to the mass extinctions of certain species.[7]

The assumption of the inferiority of Indigenous (ecological) knowledge resulted in beliefs that their practices had negative ecological implications and the subsequent adoption of colonial conservation policies that prohibited Indigenous groups of accessing traditional lands or engaging in traditional agricultural practices, measures that had significant consequences for Indigenous livelihoods and knowledge system.[8]

These momentous shifts in nature-society relations swelled with the progression of modernity, continued establishment of colonial empires throughout the non-Western world, and, very importantly, the growth of capitalism. Regarding the latter, the idea of the inferiority of nature provided a basis for the birth and expansion of capitalism, which was underpinned by the deepening authority and consolidation of the science-state-capital complex.[9] As such, in tandem with the strengthening of capitalism came a particular modern view of nature-society relations that, on the one hand, valued nature in so much as it could be used to benefit humans, and on the other hand, prioritized and centered scientific knowledge and technology for the exploitation of nature as a means of expanding capitalist relations.[10] This approach to nature-society relations legitimized the objectification, appropriation and exploitation of nature leading to its consumption and commodification in the name of progress and development, a key pillar of the modern ideology of developmentalism. In this way, as Machado Aráoz argues,[11] the semiotics of nature was transformed with the ascendance of the narrative of modernity, which, on the one hand, decoupled nature from humans and the sacred divine and, on the other hand, reframed nature through a utilitarian lens, thereby transforming its overall significance.

---

[7] Francis, Romain. "The Tyranny of the Coloniality of Nature and the Elusive Question of Justice: Challenging Discourse and Searching for Alternative Paths." In *Reimagining Justice, Human Rights, and Leadership in Africa*, 2020, 39–57. Cham, Switzerland: Springer, p. 46.

[8] Ibid.

[9] Machado Aráoz, Horacio. "La 'Naturaleza' como Objeto Colonial: Una Mirada desde la Condición Eco-Bio-Pólitica del Colonialismo Contemporáneo."

[10] Capigas-Rotundo, Juan Camilo. "The Biocoloniality of Power", in: Castro-Gómez, S; Grosfoguel R. (Ed.) 2007 El Giro Decolonial. Reflexiones para una diversidad epistémica más allá del capitalismo global [The decolonial turn. Reflections for an epistemic diversity beyond global capitalism]. Pontificia Universidad Javeriana -Instituto Pensar-. Universidad Central—IESCO-, Siglo del Hombre Editores. Bogotá.

[11] Machado Aráoz, Horacio. "La 'Naturaleza' como Objeto Colonial: Una Mirada desde la Condición Eco-Bio-Pólitica del Colonialismo Contemporáneo."

Modernity and Multi-Species Justice

The recognition of the binary nature of multi-species relations and the ensuing social relations is a core element of analyzing multi-species justice. A significant component is the recognition of 'the other' as a legitimate actor/stakeholder/participant in processes, policymaking and decision-making and this applied even more to multi-species justice. The usual starting point for such debates focuses on the concept of rights and thus much attention will be given to rights here although the connection between rights and representation is of course more complex.

In the context of environmental justice (EJ) in general, Schlosberg viewed recognition as forming the basis for both distributional and procedural justice. This opened a door for deeper analytical consideration of social and cultural hierarchies for achieving EJ, which aids in getting at the sources of inequality that distributional and procedural forms of justice are unable to address on their own. It is here that we can see how Schlosberg[12] concretely drew on the structural approach of Fraser[13] when he introduced the concept to the EJ literature. In doing so, his work points out the limitations of distribution for meeting the global environmental movement's diverse demands for justice. His interpretation of recognition saw the concept as applying to groups and communities, and not just individuals. While this concept is applied in an anthropocentric context by Schlosberg, it is equally relevant in a multi-species justice context.

In a later iteration of REJ, Schlosberg[14] incorporated the capabilities approach, drawing on the work of Martha Nussbaum and Amartya Sen,[15] into his concept of EJ. The human capabilities approach by Sen contains a normative argument for an accomplishment-based understanding of justice that goes beyond economic well-being to include concepts such as self-determination, respect and dignity. It transcends material capabilities such as food and shelter to include an accomplishments-based understanding of justice that takes into consideration 'the lives that people can actually live'.[16] Sen takes a very inclusive view of life that includes the fulfillment of

---

12 Schlosberg, David. "The Justice of Environmental Justice: Reconcling Equity, Recognition, and Participation in a Political Movement." In *Moranl and Political Reasoning in Environmental Practice*, edited by Andrew Light and Avner de Shalit. Cambridge, MA: MIT Press, 2003.

Schlosberg, David. "Reconceiving Environmental Justice: Global Movements and Political Theories." *Environmental Politics* 13, No 3 (2004), 517–40.

13 Fraser, Nancy. 2000. "Rethinking Recognition." *New Left Review* May/June: 107–120.

Fraser, Nancy. *Redistribution or Recognition*. London: Verso, 2003.

14 Schlosberg, David. *Defining Environmental Justice: Theories, Movements, and Nature.*

15 Nussbaum, Martha. *Frontiers of Justice: Disability, Nationality, Species Membership*. Cambridge, MA: Harvard University Press, 2006.

Nussbaum, Martha, and Amartya Sen. *The Quality of Life*. Oxford: Oxford University Press, 1992.

Sen, Amartya. "Capability and Well-Being." In *The Quality of Life*, edited by Martha Nussbaum and Amartya Sen. Oxford: Clarendon Press, 1992.

Sen, Amartya. *Commodities and Capabilities*. Oxford: Oxford Univesity Press, 1999a.

Sen, Amartya. *Development as Freedom*. New York: Anchor, 1999b.

16 Sen, Amartya. The idea of justice. Harvard University Press: Cambridge, 2009, p. 18.

material needs, political participation, social recognition, dignity and respect, though limits his analysis to the individual, situating this theory under the umbrella of liberal theories.[17] Schlosberg,[18] however, makes the argument that EJ centers on the individual and community ability to function in accordance with their capabilities, thus taking a more 'radical' reading of the theory than Sen's and Nussbaum's more liberal approach, given the expanded focus on communities.[19] This will be developed in a multi-species context below.

Further, Indigenous approaches to EJ are based on knowledge systems predicated on Indigenous relational ontologies that envision a broad system characterized by inter-relations and reciprocity between humans and non-humans, with the latter including animals, plant life, territory, the supernatural/spiritual world, weather elements, celestial bodies, etc.[20] In this expanded frame, humans are not the only victims of environmental injustice, but also nature and territory, and thus Indigenous discourses call for the protection of the rights of nature and territory in addition to those of humans.[21] Indigenous notions of justice are supported by systems of ancestral law and legal practices that aim to maintain balance between "all orders of beings"—human and non-human—to ensure the continuance of life.[22] At the heart of these systems are the principles of reciprocity and responsibility. As Whyte explains, "…responsibilities refer to the reciprocal (though not necessarily equal) attitudes and patterns of behaviour that are expected by and of various parties by virtue of the different roles that each may be understood to play in a relationship.[23] Reciprocity is understood through the gift-giving and -receiving relationship in which each party has a special contribution to make". Some examples of Indigenous expressions of EJ include the concepts of *Buen Vivir*, which comes out of South American Indigenous thought from the Andes and *minoo-mnaadmozawin* that emerges from Anishaabek traditions in North America.

Conceptually, a discussion on multi-species justice needs to start with a brief foray into the field of animal rights as a starting point to the discussion on rights of nature.

---

[17] Svarstad, Hanne, and Tor A. Benjaminsen. "Reading Radical Environmental Justice Through a Political Ecology Lens." *Geoforum* 108 (2020): 1–11. https://doi.org/10.1016/j.geoforum.2019.11.007.

[18] Schlosberg, David. *Defining Environmental Justice: Theories, Movements, and Nature.*

[19] Svarstad, Hanne, and Tor A. Benjaminsen. "Reading Radical Environmental Justice Through a Political Ecology Lens.".

[20] Escobar, Arturo. 2015. "Territorios de Diferencia: La Ontología Política de Los 'Derechos Al Territorio.'" [Territories of Difference: The Political Ontology of 'Rights to Territory'] *Desenvolvimento e Meio Ambiente* 35 (2015), 89–100. https://doi.org/10.5380/dma.v35i0.43541.

MacGregor, Deborah. "Indigenous Environmental Justice, Knowledge, and Law." *Kalfou* 5, No 2 (2018): 279–296. https://doi.org/10.15367/kf.v5i2.213.

Ulloa, Astrid. "Perspectives of Environmental Justice from Indigenous Peoples of Latin America: A Relational Indigenous Environmental Justice." *Environmental Justice* 10, No 6 (2017), 175–180. https://doi.org/10.1089/env.2017.0017.

[21] Ulloa, Ibid., p. 3.

[22] MacGregor, Deborah. "Indigenous Environmental Justice, Knowledge, and Law."

[23] Whyte, Kyle. "Settler Colonialism, Ecology, and Environmental Injustice." *Environment and Society: Advances in Research* 9, No 9 (2018), 125–44: 131.

Here, a distinction needs to be made between 'wild' animals and animals in captivity in a multi-species discussion. There are also subgroups between these. The animal rights literature distinguishes between 'wild' animals that live away from humans and 'liminal' animals that live among humans.[24] The ensuing ethical arguments are not directly relevant for the discussion here on coloniality, nature-society relations and representation, i.e. regarding environmental justice.

A discussion of animal rights is ultimately a discussion of nature—society relations with a specific emphasis on human-animal interaction. It is wide ranging and culturally specific. While this cultural specificity is recognized here, this study is primarily concerned with the commodification and unjust nature of these relations and wants to work through them conceptually as well as empirically.

Environment-Society Relations and Sentient Beings

The incommensurability of values between a social system based on accumulation of wealth, economic efficiency and unlimited growth is incompatible with a complex ecosystem. Martinez Alier argues that the economic system is organized as if it was not located within these specific ecological constraints.[25] It is not in 'the system's interest' to acknowledge its shortcomings as the system is exactly predestined to operate as a power and interest-maximizing unit. The current politico-economic system is powered by an understanding of nature and non-human sentient beings as subordinate parts of the economic machinery serving the global consumer despite the fact that it is precisely the politico-economic system that is subordinated to the workings of nature. The relationship is one of commodification and while there are critiques of this kind of nature-society relationship, it is generally accepted uncritically except by animal rights activists, critical academics and most environmental scientists and ecologists. However, in the decision-making centers of economics and politics, no such debate is taking place in a serious or meaningful way. One of the substantive conceptual challenges with multi-species justice is that neither nature nor particular species or indeed individual animals can represent themselves or participate in deliberative processes as they do not have the required human capabilities necessary for participation and inclusion.

Using a western frame, the environment, or nature, cannot speak for itself in human forums. It always has to be represented by someone, who then represents nature through their own particular view, interpretation or interests of/in the environment. There are some qualifying conditions to this statement.

First of all, the dichotomy of nature and society holds in hegemonic structures dominated by the western economic system, but it does not hold for most Indigenous

[24] Donaldson Sue & Will Kymlicka. "A Defense of Animal Citizens and Sovereigns", LEAP, 1 (2013): 143–160.

[25] Martinez-Alier, Joan. *The Environmentalism of the Poor; A Study of Ecological Conflicts and Valuation*. Cheltenham: Edward Elgar, 2002.

societies. Indigenous peoples are becoming more visible participants in 'mainstream' global forums and here these two worlds collide. Indigenous peoples are perceived (and self-perceive) as guardians of nature or as having such a close relationship with nature that by definition they are part of nature rather than part of a nature-society dichotomy. They are vitally important in conservation given that 22% of global land is under the stewardship of Indigenous peoples and that this land contains 80% of remaining biodiversity and 20% of forest carbon stocks.[26]

As acknowledged in the Principles and Guidelines for the Protection of the Heritage of Indigenous peoples, Indigenous knowledge is a "complete knowledge system with its own concepts of epistemology, and its own scientific and logical validity".[27] Indigenous peoples have a constantly evolving relationship with their environment and see themselves as part of this relationship rather than as the dominating element. This is a fundamentally collective rather than individualistic relationship that sees humans/human society as equal to other elements of nature rather than its master or as being in a privileged position above sentient other actors. Because of these strong contextual and cultural connections Indigenous knowledge becomes an essential part of Indigenous peoples' lives as it provides the necessary means of survival.

As Magni puts it:

> Central to Indigenous peoples' livelihood is the notion of living well, which is based on a set of norms and values shared by different Indigenous groups across the world. Fundamental to these values are aspects such as the community, the harmonious relationship between human-nature-universe and notions of equality and complementarity (Ibañez, 2010; Gudynas, 2011). Among the fundamental conditions of well-being are sufficient food, strong family and community values of caring, reciprocity and solidarity, freedom to express identity and to practice one's culture and safe and non-polluted environment (UNPFII, 2010).[28]

What distinguishes the inclusive Indigenous view and relationships with nature from the global west/north's nature-society dichotomy is first and foremost the concepts of commodification, accumulation and mastery. The dualism of nature versus society, of culture versus nature, of subject versus object, of self versus others and ultimately of civilized versus primitive determines the hierarchies of knowledge, of legitimacy, of representation and ultimately of power as it implies and legitimates authority and power. The place of sentient other-than-human beings in the global system is sealed and legitimized through these dualisms as practiced by the hegemonic system. Indigenous societies are not based on such dualism and therefore are the exception in nature-society relations exactly because they do not distinguish between nature and society as separate entities.

---

[26] Suiseeya, Kimberley, Laura Zanotti & Kate Haapala, "Navigating the Spaces Between Human Rights and Justice: Cultivating Indigenous Representation in Global Environmental Governance", *The Journal of Peasant Studies* (2021) https://doi.org/10.1080/03066150.2020.1835869.

United Nations. *State of the World's Indigenous Peoples*. New York, NY: United Nations, 2009.

[27] Battiste, Marie. *Indigenous Knowledge and Pedagogy in First Nation Education*, National Working Group on Education and the Minister of Indian Affairs, Ottawa CN, 2002, p. 7.

[28] Magni, Giorgia. Indigenous knowledge and implications for the sustainable development agenda, background paper prepared for the 2016 Global Education Monitoring Report, 2016, p. 7.

This has some important implications. Indigenous peoples have become increasingly visible actors in global environmental and climate change forums and have quietly and gradually successfully been pushing for the incorporation of rights for nature in global treaties. We will return to this important issue later in this chapter in more detail. This bestowing of rights can be found, for example in the constitutions of Ecuador and Bolivia (through buen vivir), through tribal law in the United States or the city of Curridabat in Costa Rica that gives citizenship rights to local pollinators.[29] Such success has not been reproduced at the global level despite the increasing participation of Indigenous groups (though of course participation does not mean decision making rights).

Before engaging with legally recoverable rights, it is necessary to engage with the ethics of animal rights in general. There have been a range of recent publications on this subject which constitute a veritable 'animal turn'. These which will be briefly reviewed.

The Ethics of Animal Rights

With the rising awareness regarding the lamentable state of animal welfare in factory farms, with the haunting deaths of wildlife in catastrophic events associated with climate change such as the Australian wildfires or the Californian fires and the general rise of vegetarian and vegan lifestyles, the question of animal rights, ethics and justice has become a much-debated issue in many academic fields as well as in the general public sphere. By this is meant a much wider concept of animal rights than the one traditionally and historically associated with animal rights activism in the 1990s and beyond. In this section, I will briefly review the main arguments.

A starting point for the concept is situated in the field of political philosophy. Cochrane refers to these evolving debates as the political turn in animal ethics.[30] As in environmental justice discourses in general, much of the literature is grounded in liberal thought. Cochrane's argument is based on marrying animal ethics and cosmopolitan democracy (as opposed to more relational, differentiated approaches discussed below). Theories of animal rights have only recently started to develop political accounts of what the human-animal relationship should entail. Until recently, the animal ethics literature was centered on what moral obligations humans have toward animals rather than on what political structures should exist given the all-pervasive human impact on other sentient beings.[31]

The first question that bestows legitimacy to this kind of inquiry is whether being sentient means one should have rights or whether those two issues are decoupled.

---

[29] Greenfield, Patrick, Sweet City: The Costa Rica Suburb That Gave Citizenship to Plants, Bees and Trees, *The Guardian*, 29 April 2020 (https://www.theguardian.com/environment/2020/apr/29/sweet-city-the-costa-rica-suburb-that-gave-citizenship-to-bees-plants-and-trees-aoe).

[30] Cochrane, Alasdair. *Sentienist Politics*. Oxford: Oxford University Press, 2018.

[31] Ibid., p 4.

Most political philosophy, the basis of all discussion on the subject of who has and ought to have rights, has long been silent on the subject of rights for other-than-human beings just as it was long silent on the rights of various human groups that have only recently been given political rights in the past 100 years or so. Berkey argues that, following Rawlsian interpretations of justice, entitlements to justice or direct representation apply only to those who are contributing to, or are capable of contributing to, cooperative schemes for mutual advantage as exemplified by modern societies.[32] As non-humans are generally not believed capable of making such contributions, this canon of the literature then excludes them from direct entitlements to justice. So, just as in the environmental justice debates, Rawls and liberal concepts of rights are very selective in who has rights and who has the right to be represented.

Other writers such as for example Martha Nussbaum depart from the Rawlsian concept of justice and position entitlements to justice in the interests that beings have in living dignified and flourishing lives. As this interest is shared by humans and animals alike, that gives them direct entitlements to justice.[33] Based on her capabilities approach, Nussbaum's entitlements to justice are based on moral and compassionate claims which distinguishes her approach from the more rigid Rawlsian approach which separates justice and morality. The Rawlsian basis for justice necessitates an active actor as an object and subject of justice while Nussbaum's capabilities approach would give justice entitlement based on moral grounds which could be incorporated into legal and political frameworks without active agency of the animal, or group of animals, in question. Likewise, Pellow argues for more intersectionality and the inclusion of the more than human world when legitimizing and recognizing actors and concepts.[34] In the same vein, Indigenous concepts of organization, justice and social relations fully incorporate the 'non-human- world' as equal actors, participants and stakeholders.

Meijer feels that the crux of the argument is that animals are perceived to be as not capable of communication and thus not capable of political action. Therefore, they cannot be accommodated in the Rawlsian framework of justice.[35] However, Meijer argues that animals are known to be able to communicate. The natural sciences show convincingly that animals communicate with each other, and she refers to Donaldson and Kymlicka and their political theory on animal rights giving legitimacy to animal communication as proof of the moral imperative of including animals as political actors. Hinchcliffe, on the other hand, uses the concept of personhood as his starting point. He argues that.

---

32 Berkey, Brian. "Prospects for an Inclusive Theory of Justice: The Case of Non-Human Animals," *Journal of Applied Philosophy*, 34, No. 5 (2017): 679–695, p. 679.

33 Nussbaum, Martha. *Frontiers of Justice: Disability, Nationality, Species Membership*. Cambridge, MA: Harvard University Press, 2006, p 390.

34 Pellow, David Naguib. *What Is Critical Environmental Justice?* Cambridge, UK: Polity Press, 2018, pp. 17/18.

35 Meijer, Eva. "Political communication with animals", ANIMaLIA, 5, No 1 (2013): 28–52.

Nussbaum, Martha. *Women and Human Development: The Capabilities Approach*. Oxford: Oxford University Press, 2000.

> Traditionally, animal rights theory focuses on establishing the basic moral status of animals, accompanied by the same kind of universal inviolable moral rights, e.g., against torture and murder, that we associate with personhood in the case of humans. In the human case, the rights of personhood are often thought to be grounded in "characteristically human" traits, such as a capacity to reason or possession of a subjective experience or sentience. By arguing that certain animals also possess these characteristics to a sufficient degree, animal rights theorists have tried to show that such animals also deserve to be recognized as persons.[36]

However, these authors also argue that as a result of this inclusion, the concept of personhood, or justice, or rights, or citizenship (depending on which angle the animal rights writers approach the subject from) would need to be amended to include the different kind of relations that would exist in a polis that includes both animals and humans as political actors.

Cochrane notes that EU legislation recognizes animals as sentient beings and recognizes that moral obligations arise out of that recognition.[37] However, he feels that this provision does not constrain political action or fundamentally change human-animal relations. While there surely are some animal welfare policies resulting out of this recognition, it does not put a stop to egregious practices in factory farming, for example, and seems to be perfectly compatible with it. Cochrane takes his argument further and argues that while the "worth and rights of humans are conventionally understood to provide the justification for the exercise of political power, indeed, a state which systematically violates human rights loses its moral authority".[38] What naturally follows is that if other sentient beings also have worth and rights, then the aim and structure of the political sphere would need to be constructed around that recognition and a different political system would need to evolve as a result.

However, before engaging with that context, it is necessary to go over Donaldson and Kymlicka's[39] argument in greater detail and explore its basic tenets and consequences. Donaldson and Kymlicka argue that the debates on animal rights have been at a standstill for a while (though their book certainly kick started the debates again). Their main argument is that we as society must move past basic and universal rights for animals and explore the normative consequences of giving them member status of our political communities. Thus, they argue, the concept of citizenship should be redefined to include animals.

Donaldson and Kymlicka base their argument on three basic assumptions. First, they see the key purpose of justice to protect vulnerable individuals.[40] Second, both animals and humans are sentient. This sentience gives both humans and animals the vulnerability that justice is supposed to protect. As a result, third, animals have

---

[36] Hinchcliffe, Christopher. "Animals and the limits of citizenship: zoopolis and the concept of citizenship", *Journal of Political Philosophy*, 23, No 3 (2015), 302–320, p. 304.

[37] Cochrane, Alasdair. "Cosmozoopolis, the Case Against Group-Differentiated Animal Rights", LEAP, 1 (2013), 127–141.

[38] Cochrane, Alasdair. *Sentienist Politics*, 2018, p. 3.

[39] Donaldson, Sue & Will Kymlicka. *Zoopolis, a Political Theory of Animal Rights*. Oxford: Oxford University Press, 2011.

[40] Ibid., p. 33.

certain basic inviolable rights just like humans do.[41] Following from this, if both animals and humans are interest-bearing subjects, then human-animal relations are intersubjective, they are relations between selves. Thus, they can be relations based on obligations and rights. Donaldson and Kymlicka focus their argument on pets and what they call 'liminal' animals—those animals that live near human communities such as birds, rodents, raccoons. They do not address factory farm animals in their argument or farm animals in general (nor 'wild' animals for that matter) which creates a hierarchy of animals as an unintended consequence, in fact a zoological class-based society. The implication being that certain animals are more worthy of rights than others based on their proximity to humans. Ultimately, it also means that while all humans have rights under this line of argument, only some animals have rights which establishes a hierarchy of sentient actors or only attributes sentience to liminal animals.

According to Donaldson and Kymlicka, in addition to our negative duties toward animals (such as not to kill or torture or otherwise mistreat), humans could also have positive duties toward animals: those duties that come with treating them as co-members of our political communities. And it is exactly this political component that Donaldson and Kymlicka feel has been neglected by traditional animal rights ethics so far. To be clear, this should include acknowledging that they are not another category of humans and both citizenship theory and animal rights theory must acknowledge and incorporate that.

The first problem or obstacle to this view is that it is hard to perceive of animals having political agency. While it is possible to see animals as a group that has rights bestowed upon them by human society, these are passive rights that are not actively enforceable by animals themselves for a variety of reasons—political agency, consciousness of these rights, ability to negotiate these rights, etc. But Donaldson and Kymlicka argue that this is an unduly narrow focus on the concept of citizenship and that the capacity to represent oneself in the political process is not necessarily a condition for citizenship (after all, children and cognitively challenged members of the community also have citizenship).[42] Citizenship is not just about political agency but also about being a citizen of a state (in terms of being a passport holder) which is a much more narrow, passive form of citizenship. Thus, they reject the view that animals are incapable of citizenship because of a lack of political agency.

Nevertheless, while Donaldson and Kymlicka see the comparison with children and cognitively disabled members of society with animals as valid, surely children may be citizens in the sense of holding a nationality but they are not given rights to participate in the political process as such. By the same token, the rights of the cognitively disabled are largely symbolic. The rights of children and cognitively impaired humans are usually represented by their legal guardians. Thus the same would apply inter alia to animals. But in many ways, this would be akin to the legal

[41] Ibid., Chapter 2.

[42] Donaldson Sue & Will Kymlicka. "A Defense of Animal Citizens and Sovereigns," LEAP, 1 (2013),143–160.

definitions of rights rather than the political concept of rights, which, after all, is based on active participation in the political process.

Further, Donaldson and Kymlicka also discuss some specific rights possessed by domesticated animals such as the right to protection from predation (presumably from predation by humans rather than fellow animals as that would not be compatible with the way food chains work). The right to protection from predation is effectively a corollary to the right to life following Cochrane.[43] Surely that argument in itself would commit humanity to vegetarianism and put an end to factory farming immediately if animal rights extended beyond domestic and liminal animals (which it does not, in Donaldson and Kymlicka's argument).

But what is the most problematic aspect of Donaldson and Kymlicka's argument is that it effectively establishes a hierarchy, or a taxonomy of value from an anthropocentric perspective. This is compatible with their concept of citizenship (based on a concept of 'sovereignty' that includes politically disenfranchised members of society such as children, the cognitively disabled but also slaves) and effectively establishes classes of citizenship and uncritically takes this as valid. As Cochrane argues:

> One way in which outsiders may be denied their just entitlements is through privileging the rights of "insider" groups. Granting too much weight to the rights of those close to us, after all, leaves fewer resources to secure the rights of those further away...Donaldson and Kymlicka argue that the right to healthcare is a basic right of citizenship in most contemporary societies, and is a right that should be extended to domesticated animals. Crucially, they also believe that it is a right that should not be attributed to wild animals (142-143)... My point is rather to show that Donaldson and Kymlicka's privileging of the rights of the group "domesticated animals" is illegitimate.[44]

It is uncontroversial that animals cannot be active participants in political processes because they lack the ability to represent themselves. Many political theorists do not see that as a problem while others do. But ultimately it means that animals as sentient actors will have to be represented by someone who speaks on their behalf in western forms of political organization. The question, then, rather, is whether that can count as having citizenship. Under Donaldson and Kymlickas's view it does count. However, under Cochrane or Nussbaum it does not—but neither is the obtainment of citizenship a necessity for multi-species justice under their frameworks. Nevertheless, there have been quite a few recent legal advances that have given nature (and thus animals) legal rights without resorting to deliberations on citizenship. These will be explored below.

---

[43] Ibid.

[44] Cochrane, Alasdair. "Cosmozoopolis, the case against group-differentiated animal rights," LEAP, 1 (2013), 127–141, p. 134.

Legal Attempts to Give Nature Rights

In February 2019, the citizens of Toledo/Ohio, in a referendum, voted to grant the Lake Erie ecosystem the legally enforceable right 'to exist, flourish and to naturally evolve', thus establishing the Lake Erie Bill of Rights.[45] The Lake Erie Bill of Rights (LEBOR) would have given citizens the right to sue polluters on behalf of the lake in the same way nature has rights in tribal courts in the United States. However, in July 2019 Governor Mike DeVine signed a law stating that ecosystems cannot have legal rights under Ohio law, thus delivering a fatal blow to LEBOR.

In April 2020, the Guardian ran an article in its 'Wild Cities' series on the San Jose/Costa Rice suburb of Curridabat which has granted citizenship to bees, bats, hummingbirds and butterflies as the key pollinators and has created biocorridors throughout the suburb.[46] While this honorary citizenship does not bestow any legal rights, it is part of a wider movement in Latin America, and particularly Costa Rica, that recognizes the importance of wildlife for a healthy human life. Curridabat actively creates habitats for pollinators.

The constitutions of Ecuador and Bolivia include the concept of buen vivir which acknowledges the nature-society relations of Indigenous peoples and have enshrined constitutional rights for 'pachamama', or Mother Earth. These rights have been tested in court and upheld.[47] The concept of buen vivir is generally seen as the first attempt to institutionalize rights for nature in a binding legal framework.

All around the world rivers have been granted legal personhood such as in the cases of the Whanganui river in New Zealand in 2017 and the rivers Ganges and Yamuna.[48] A similar process is in motion for the river Ouse in East Sussex/UK.[49] Indigenous leader in the South Pacific have signed a treaty to give whales legal personhood.[50] All these efforts, some successful and some not, some legally enforceable and some symbolic, come together to indicate that there is a larger global movement working toward the concept of giving nature citizen or personhood rights.

---

45 Berman, Devon. "Lake Erie Bill of Rights gets the Ax: Is Legal Personhood for Nature dead in the water?" *Sustainable Development Law and Policy*, 20, No. 1 (2020), 15/16.

46 Greenfield, Patrick, Sweet City: the Costa Rica suburb that gave citizenship to plants, bees and trees, The Guardian, 29 April 2020 (https://www.theguardian.com/environment/2020/apr/29/sweet-city-the-costa-rica-suburb-that-gave-citizenship-to-bees-plants-and-trees-aoe).

47 Berman, Devon. "Lake Erie Bill of Rights gets the Ax".

48 Ibid.

49 Kaminski, Isabella, The river Ouse may become first in England to gain legal rights, *The Guardian*, 1 March 2023 (https://www.theguardian.com/environment/2023/mar/01/sussex-river-ouse-first-in-england-legal-rights-aoe?utm_term=64002d5ca118ca5030b87e16a1389fae&utm_campaign=GuardianTodayUK&utm_source=esp&utm_medium=Email&CMP=GTUK_email).

50 https://www.npr.org/2024/04/03/1242451885/whale-personhood-climate-change-polynesia-maori-new-zealand

The United Kingdom is actively working on a set of bills that recognize animals as sentient beings.[51] It will cover farm animals and pets in the UK as well as be applied to imports of, for example, shark fins or animal trophies. While the debate on sentience and the debate on rights of nature per se, are conducted in separate realms, I consider there to be a strong overlap between the two as they both grapple with multi-species justice.

What is the difference between citizenship and personhood in regard to multi-species justice? One is a political concept, and the other is a legal concept to start with. Citizenship comes with moral obligations. The concept of legal personhood goes beyond individuals and applies to corporations and other legal bodies such as unions or federations. Thus, rather than bestowing political rights, the concept gives standing in court. This concept of standing, i.e. the right to sue (and be sued as well of course) has long been contested in relation to nature, going back to 1972 and Christopher Stone's seminal article *Should Trees Have Standing.*

Mancilla summarizes Stone's argument as follows:

> In Trees, Stone enumerates the three necessary and sufficient conditions for an object to have legal standing, i.e., to initiate an action in court or, more precisely, to institute judicial review. First, it must be able to institute actions at its behest (this does not mean that they have to speak for themselves, but merely that they can have someone speak for them, as is also the case of states, infants, incompetents, corporations, universities, etc.). Secondly, in determining the granting of legal relief, the court must take direct injuries to it into account. Thirdly, the relief must be to the benefit of it; for example, by making it 'whole' again. Insofar as these three conditions are met by natural objects such as forests, rivers or species, their assigned representatives or guardians should be able to file suit when they consider that their interests might be adversely affected.[52]

Stone's intention was to engage a debate that rights, rather than just protective laws, rules and norms, give flexibility and an open-endedness that would not otherwise exist. In other words, a debate on nature's rights is a debate about consciousness and a critical questioning that the mere establishment of laws and norms within an existing paradigm would not achieve.

Garthoff argues that having legal standing is usually associated with having liability.[53] But "important problems with the standard understanding of legal personhood stem from the fact that the capacities needed to possess rights (standing) are not the same as the capacities needed to possess obligations (liability)". Humans have obligations because they possess reason and can therefore moderate their actions and behavior. Humans are distinct from non-human animals in that they have the faculties for critical reasoning from which follow moral and legal obligations in our society. Many non-human animals are also conscious. For example, many pets seem

---

[51] Harvey, Fiona, Animals to be formally recognized as sentient beings in UK Law, *The Guardian*, 12 May 2021, https://www.theguardian.com/world/2021/may/12/animals-to-be-formally-recognised-as-sentient-beings-in-uk-law?utm_term=7198793722e9e53e8e9029ed42411231&utm_campaign=GuardianTodayUK&utm_source=esp&utm_medium=Email&CMP=GTUK_email.

[52] Mancilla, Alejandra. "Should Trees Have Standing? by Christopher Stone". In *Pandora's Box*, 18 (2011), 91–99, p. 92.

[53] Garthoff, J. (2018): https://scholars.org/brief/corporations-animals-and-legal-personhood.

to feel guilt after doing something that is not "allowed". This class of conscious but non-rational extends to some humans such as fetuses, infants and the severely cognitively impaired. Garthoff contends that "non-rational beings with conscious experiences have moral standing".[54] Because of this moral standing, society cannot disregard them without moral fault. In recognition of these facts, these individuals should have legal standing even when they lack legal personhood.

Thus, from a legal perspective, there is a persuasive normative argument to be made that makes a case for extending legal rights to animals and to nature in a way that the liberal political theory arguments cannot present. Unfortunately, the US Supreme Court's narrow interpretation of legal standing makes it impossible to advance this case in the United States though, as the above examples show, there is more hope for a paradigm shift elsewhere or in global governance.

Multi-Species Justice in the Anthropocene—Concluding Thoughts

The commodification of other-than-human species is a vital component of the construct that provides the scaffolding in which the Anthropocene evolved. The roots of modern multi-species relations can be found in Enlightenment thinking which provides the theoretical basis of modernity and relegates other-than-human beings in a binomic ordering. More recent political thought on other-than-human beings has provided different avenues of thought that make the argument for a rights-based approach to multi-species relations, aiming to lead to more multi-species justice.

What this debate on the rights of nature and on sentience has shown is that our current understanding of nature-society relations lacks recognition for other species as 'worthy' of representation. Although rights-based approaches emulate the liberal, individualistic roots of a binomic system, they are nevertheless the most promising avenue for multi-species representation in the Anthropocene and would be one way forward to mitigate some of the most problematic aspects of the Anthropocene by questioning multi-species relations in their current form.

---

[54] Ibid.

Chapter 5
The Capitalocene as a Kinocene: Unveiling Environmental and Political Dualisms Subjecting Migrations in the World-Ecology Production

Yoan Molinero-Gerbeau and Victor Pérez-Segura

Introduction

The intensity of the contemporary climate emergency has triggered a historically unprecedented environmental awareness in the social sciences. The rise of green theoretical currents, research fields such as Anthropocene Studies, and interdisciplinary debates on the impact this has on phenomena like migration are good examples of this.[1]

While transcending the "ecological blindness"[2] that has characterised the research community is a positive development in the search for ways out of climate change, the positions that have emerged reveal the existence of an intense dispute over "narratives". The liberal camp, hegemonic thanks to the support that States, International Organisations and other actors exercising important quotas of power have given it, has taken on two concepts of notable importance: Sustainability and the Anthropocene. The first refers to the backbone and cross-cutting promotion of green capitalism,

[1] Yoan Molinero-Gerbeau, *El Medioambiente En Las Relaciones Internacionales* [The Environment in International Relations] (Madrid: Síntesis, 2022).

[2] Karlos Pérez de Armiño, "La Teoría Verde En Las Relaciones Internacionales. Aportes y Desafíos de Una Corriente Emergente" [Green Theory in International Relations: Contributions and Challenges of an Emerging Current] in *100 Años de Relaciones Internacionales. Una Mirada Reflexiva* [100 years of International Relations. A Reflexive view] eds. Caterina García Segura, José Antonio Sanahuja and Francisco J. Verdes-Montenegro (Valencia: Tirant Lo Blanch, 2020): 337–366.

Y. Molinero-Gerbeau (✉) · V. Pérez-Segura
Instituto Universitario de Estudios sobre Migraciones, Universidad Pontificia Comillas, Madrid, Spain
e-mail: ymolinero@comillas.edu

V. Pérez-Segura
e-mail: vperezs@comillas.edu

M. Robson (ed.), *The Anthropocene, Ontopolitics, and International Relations*,
https://doi.org/10.1007/978-981-96-2925-1_5

perfectly reflected in the 2030 Agenda. The second, on the other hand, is based on the premise that the rise in global temperatures would mean the end of the Holocene, ushering in a new geological period driven by humanity as a whole.

Migration, as one of the most notable social phenomena of contemporary times, has also been permeated by both perspectives, with global hegemonic powers promoting new fragmentations and labelling to characterise human mobility. According to these, anthropogenic climate change would generate new mobilities, naming the supposed subjects who would undertake them with a myriad of categories such as "climate refugees" or "environmental migrants".[3]

Starting from a deconstruction of the hegemonic Anthropocene perspective, this chapter will argue for placing contemporary migrations within the framework of the Capitalocene, a perspective locating the responsibility for climate change in global capitalism. We will show how the mobility of human beings is intrinsically linked to the processes of accumulation, which is why we can speak of a Kinocene, a label that, although Thomas Nail has espoused it, we will also dismantle to place it around other parameters. The aim is none other than to show how contemporary migrations are deliberately fragmented around dualisms that seek to obscure both the real bases of these mobilities and to reduce their autonomy by favouring their control by states. In doing so, we will attempt to reposition the debates around the climate crisis, and more specifically around migrations, within parameters that allow them to be "liberated"[4] from the subjection in which the hegemonic actors of the world-ecology want to keep them.

To this end, this chapter will start addressing the debate around the Capitalocene by locating our proposed sub-perspective or "history" of the Capitalocene as the Kinocene. The next section will discuss the structural factors linking human mobility to the Kinocene driven by the capitalist world-ecology. Following this, the third section will delve into how hegemonic perspectives have articulated two dualisms to bring migration under their control. Finally, the chapter will close with a series of conclusions aimed at proposing a cognitive break with these dualisms to articulate frameworks that not only allow for a holistic understanding of the migration phenomenon but also contribute to its autonomisation from state apparatuses.

---

[3] Victor Pérez-Segura, "*Horizontes y Fronteras de Los Estudios Sobre Las Migraciones Climáticas. Una Aproximación Meta-Teórica a Los Desplazamientos Asociados a La Degradación Medioambiental*" [Horizons and Frontiers of Climate Migration Studies: A Meta-Theoretical Approach to Displacements Associated with Environmental Degradation] (PhD diss., Universidad Pontificia Comillas, 2024).

[4] Gennaro Avallone and Yoan Molinero-Gerbeau 'Freeing Migration: The Contribution of Abdelmalek Sayad to a Migrant-Centric Epistemology', *Migraciones Internacionales* 12 (2021): 1–26.

The Capitalocene as a Kinocene or the Kinocene as a Capitalocene "History"

As the introduction to this book has already addressed, the field of what we can call "Anthropocene Studies" is vibrant and constantly changing.[5] A particular fact about it, and one that sets it apart from other areas of study, is how prolific it has been in coining neologisms intended to characterise the current geological period.[6] Thus, rather than merely harbouring a classic debate between theoretical positions, a particular feature of these studies is that a multiplicity of authors have proposed using new labels that better reflect both the defining features and the origins and characteristics of the contemporary climatic period. Some authors say this is chaotic and unhelpful, thus affecting the understanding of climate emergency.[7] Still, for others, including the authors of this chapter, it is the opposite, having been considered an extremely enriching exercise.

Anyway, this does not imply that the proliferation of "-cenes" does not entail some problems, such as the articulation of insufficiently developed theoretical perspectives or the use of labels that seem more like attempts by their authors to claim academic success and try to accumulate citations than an exercise in coining true concepts of academic value.

Without entering into this debate, Bonneuil and Fressoz[8] indirectly proposed an extremely useful organisational criterion to arrange this confusing field of study. Starting from the Anthropocene, a framework they accept, although they develop it critically, these authors indicated the need to complexify the Anthropocene macro-framework with specific sub-narratives, which they called "Anthropocene histories". Thus, for example, they indicated that the Anthropocene is also a Thermocene or a Thanatocene to claim the role that, in the case of the former, the history of energy or, in the case of the latter, wars have played in shaping climate change.

Following this model, we consider only two sufficiently well-armed theoretical blocks (each with specific developments and undercurrents) to be considered true explanatory theories of the contemporary geological period: the Anthropocene and the Capitalocene. All other "-cenes" either fit into one of these two frameworks or do not constitute a contribution of sufficient scope to be considered as having theoretical validity.

---

[5] Thomas Hickmann, Lena Partzsch, Philipp Pattberg, and Sabine Weiland, *The Anthropocene Debate and Political Science* (London: Routledge, 2018).

[6] Yoan Molinero-Gerbeau, "Antropoceno," [Anthropocene] in *El Medioambiente En Las Relaciones Internacionales*, [The environment in International Relations] ed. Yoan Molinero-Gerbeau (Madrid: Síntesis, 2022), 37–48.

[7] Carles, Soriano, 'Anthropocene, Capitalocene, and Other '-Cenes': Why a Correct Understanding of Marx's Theory of Value Is Necessary to Leave the Planetary Crisis', *Monthly Review* 74, no. 6 (2022).

[8] Christophe Bonneuil and Jean-Baptiste Fressoz. *The Shock of the Anthropocene: The Earth, History, and Us* (New York: Verso, 2016).

Since both fields are well known and this collective work gives a remarkable account of their postulates, we will limit ourselves to a superficial characterisation of both to allow a better understanding of our theoretical anchorage.

On the one hand, the Anthropocene is based on the original idea of Paul Crutzen,[9] according to which the current geological period, the Holocene, characterised by a temperate and stable climate that has allowed the emergence of human civilisations, would have ended due to human action on the environment. Thus, implementing the industrial model and the massive consumption of fossil fuels would have caused such an emission of greenhouse gases on the planet that the climate would have already mutated towards a warmer one.

The fact itself, i.e. that climate change has an anthropogenic origin, is a reality on which there seems to be unanimity (even in the Capitalocene field). This is what Moore[10] refers to as the "Geological Anthropocene", a natural science perspective that links human actions to the ecological deterioration of the planet through measurable elements.

The contested field is, however, located in what Moore termed the "Popular Anthropocene", referring to "a way of thinking the origins and evolution of modern ecological crisis".[11] Here, perspectives differ markedly, varying between, for example, those who, accepting the Anthropocene label, argue that it began with the Industrial Revolution and those who argue that it began with the "Golden peak", the time of greatest urbanisation and booming global emissions that began after the Second World War.[12]

Bonneuil and Fressoz[13] place themselves in this theoretical camp, advocating the usefulness of the Anthropocene concept but developing rather critical postulates. Indeed, it should be noted that adopting the Anthropocene framework does not imply that those who do so reproduce a hegemonic or liberal perspective, as the critical field within the Anthropocene bloc is very broad.[14]

The Capitalocene, however, differs from the critical positions of the Anthropocene field (although it shares some points with them) by setting itself up not as an explanation of human impact on climate but as a theory for understanding how the global capitalist system has historically been responsible for transforming the planetary

---

[9] Paul Crutzen, "The 'Anthropocene'", in *Earth System Science in the Anthropocene*, ed. Eckhart Ehlers and Thomas Krafft (Berlin: Springer, 2006), 13–18.

[10] Jason W. Moore, 'Name the System! Anthropocenes & the Capitalocene Alternative', World-Ecological Imaginations: Power and Production in the Web of Life, 2016, accessed July 17, 2024, https://jasonwmoore.wordpress.com/tag/anthropocene/.

[11] Moore, "Name the System".

[12] Emanuele Leonardi and Alessandro Barbero, "Il Sintomo-Antropocene," [The Symptom-Anthropocene] in *Antropocene o Capitalocene Scenari Di Ecologia-Mondo Nella Crisi Planetaria*, [Anthropocene or Capitalocene World-Ecology Scenarios in the Planetary Crisis] ed. Jason W. Moore (Verona: Ombre Corte, 2017), 7–25.

[13] Bonneuil and Fressoz, Shock of the Anthropocene.

[14] Nathanaël Wallenhorst, *A Critical Theory for the Anthropocene* (Cham: Springer, 2023).

ecology and inducing climate change. Thus, in contrast to Anthropocenists, Capitalocenists do not start from merely ecocentric analyses (considering the whole planetary biology as a structure), nor do they attribute climate change to humanity as a bloc, but, locating themselves in Structuralist analyses, they link climate change to the global capitalist system's construction of a specific ecology geared towards guaranteeing processes of accumulation.[15]

Thus, the Capitalocene does not seek to explain how particular phenomena, such as pollution, have driven climate change but rather how a capitalist world-ecology has been constructed whose appropriation of human and extra-human natures has shaped the global environment in ways that have altered the Holocene climate.

Other perspectives, including the Chtulucene[16] or the Westernocene,[17] stand as sub-narratives or "histories" (in Bonneuil and Fressoz's terms) of the Capitalocene, either because they directly draw on its framework to articulate their proposition or because they fit (with great nuance in some cases) into its theoretical corpus.

While in recent years, some studies have drawn on socio-ecological frameworks, such as world-ecology, to analyse international migration,[18,19] very little attention has been paid to the links between mobility (of both humans and the rest of nature) and climate change. Of particular interest in this regard is the perspective that Thomas Nail[20] decided to advance by proposing the "Kinocene" theory.

Starting from a "kinetic", i.e. movement-centred, framework, Nail links the increasing and intensified circulation of goods and people with an acceleration of global entropy that has broken with Holocene stability to give rise to the "age of movement".[21] In reality, for this author, this would mean not only a new historical phase but, to a certain extent, a return to the nomadic origins of humanity, whose stable settlement in certain territories, besides not being total (movement has always existed), would constitute only an exceptional phase within the broad history of mobility of our species.

---

[15] Jason W. Moore, 'The Capitalocene, Part I: On the Nature and Origins of Our Ecological Crisis', The Journal of Peasant Studies 44, no.3: 594–630.

[16] Donna J. Haraway, 'Staying with the Trouble: Anthropocene, Capitalocene, Chtulucene', in *Antropocene or Capitalocene? Nature, History, and the Crisis of Capitalism*, ed. Jason W. Moore (Oakland: PM Press, 2017), 34–77.

[17] Álvaro San Román and Yoan Molinero-Gerbeau, 'Anthropocene, Capitalocene or Westernocene? On the Ideological Foundations of the Current Climate Crisis', *Capitalism Nature Socialism* 34, no. 4 (2023): 39–57.

[18] Yoan Molinero Gerbeau and Gennaro Avallone, 'Producing Cheap Food and Labour: Migrations and Agriculture in the Capitalistic World-Ecology', *Social Change Review* 14, no. 2 (2016): 121–148.

[19] Emmanuelle Hellio and Juana Moreno Nieto, "La Ecología-Mundo Bajo Plástico: Un Análisis de La Articulación Entre La Explotación de La Naturaleza, El Racismo y El Sexismo En La Producción de Frutos Rojos de Huelva," [The World-Ecology Under Plastic: An Analysis of the Connection Between the Exploitation of Nature, Racism, and Sexism in the Production of Red Fruits in Huelva] *Relaciones Internacionales* 47 (2021): 125–142.

[20] Thomas Nail, 'Forum 1: Migrant Climate in the Kinocene', *Mobilities* 14, no.3 (2019): 375–380.

[21] Nail, Kinocene: 375.

We agree with Nail in the link he makes between certain factors such as technological development and, above all, the processes of colonisation, racism as a logic of government and, in general, the induction by core states of processes of subalternisation of the populations of the global periphery, as structural elements inducing migrations. However, we consider his theoretical argumentation of the Kinocene to be inconsistent.

By deliberately stating that "the Anthropocene and the Capitalocene are only subcategories of a much larger kinetic transformation of the Earth",[22] Nail claims that his movement theory is above both frameworks and, therefore, has a superior explanatory capacity to understand the ecological reality of the planet.

In a way, we understand Nail's approaches to be contradictory to each other. The author indicates that "Humans might have initiated this increase in movement (and capitalism certainly hastened it), but now the whole planet is producing positive feedback cycles (carbon cycles, nitrogen cycles, etc.) that have lives of their own"[23] but, at the same time, he indicates "that society has always been on the move together with nature" and that "the Kinocene shows us that there is no natural stable state of nature and no natural stable social state".[24]

In short, Nail seems to suggest that the Kinocene has always been there, at least as long as humanity has existed (he clearly reproduces Cartesian dualism), only that its cycles of movement have varied according to factors such as the rise of capitalism itself. Thus, the Capitalocene would be a phase of the Kinocene. But if this perspective assumes that the Earth has always been in motion, then the very category of the Kinocene would constitute an oxymoron since the suffix "-cene" comes from the Greek *kainos*, which means "new", when we would be talking about a supposed holistic framework characterised by its permanence over time and not by being a novelty.

In addition to this contradiction, Nail reveals his ignorance of the workings of the capitalist system by allowing himself the licence (in a somewhat sardonic tone) to assert that "if climate change did not exist, it would be necessary for capitalism to create it".[25] His reasoning for this is that climate change, by expelling populations from their territories, for example, creates new opportunities to exploit the resources present there while at the same time increasing the global reserve army.

Thus, in contrast to the classic perspectives (ranging from the liberal to the structuralist arc) that start from the premise that climate change is the result of the aggression of nature that capitalism has caused as an adverse side effect[26] or non-dualist perspectives, such as world-ecology, which consider that climate change is the result

---

[22] Nail, Kinocene: 375.

[23] Nail, Kinocene: 375.

[24] Nail, Kinocene: 375.

[25] Nail, Kinocene: 378.

[26] See, for example, Immanuel Wallerstein, 'Ecology and Capitalist Costs of Production: No Exit', in *Ecology and the World-System*, eds. Walter L. Goldfrank, David Goodman, and Andrew Szasz, (Westport: Greenwood Press, 1999), 3–12.

of the construction of a capitalist planetary ecology,[27] Nail would take a different point of view. For him, capitalism has not deliberately driven climate change but would be found to be an unexpected outcome that benefits from it significantly by opening up new commodity frontiers.

Within the field of political ecology, the conclusion of O'Connor[28] that climate change constituted "the second great contradiction of capitalism" is well known as the exploitation of the environment and natural resources for capitalist production generates an ecological crisis that, in turn, threatens the material basis necessary for the very continuation of the capitalist system. In a way, it is quite simple to understand why climate change does not benefit capitalism: being a system based on the infinite accumulation of capital, it can only deplete the bases on which it feeds, thus exhausting its sustenance in the medium and long term. In fact, as Moore[29] states, it will not be necessary to wait for the last tree to be cut down for capitalism to fall (as the famous environmentalist slogan goes) since the system will cease to sustain itself when the rate of profit begins to collapse as it no longer has sufficient resources to sustain expansive phases.

In short, we understand that the idea of the Kinocene, as Nail has put it, lacks coherence and theoretical robustness to be considered a paradigm of explanatory value for the current ecological crisis. We also consider it problematic to situate the epistemological axis on an abstract idea of "movement" without starting from the systemic dynamics and the actors that activate it. Nevertheless, we consider that capitalism-driven movements (of goods, persons, etc.) have played a key role in the systems' functioning as explanative factors in understanding the shaping of the Capitalocene. We then propose to take this label but make a re-signification to understand the Kinocene as a "history" of the Capitalocene.

Undoubtedly, the intensification of the movement of goods, people and other natures not only has an impact on the environment (with increasing carbon emissions, for instance), but the advance of the Capitalocene has certainly activated some of these movements. Climate change impacts both human and extra-human natures in structural ways, either forcing displacements, activating mobilities for productive needs (for instance, the need for peripheral workers in the global core economies) or for the appropriation of resources that all have feedback effects in deepening the Capitalocene. In the next section, we will delve into the human-nature part of these mobilities, showing what dynamics have driven and under what parameters human mobilities have been configured in the Capitalocene, making it possible to highlight its fundamental link with a capitalist *kainos* allowing us to partly elaborate a Kinocene history of the Capitalocene. The following section will delve into this structural relation.

---

[27] Jason W. Moore, 'World Accumulation and Planetary Life, or, Why Capitalism Will Not Survive Until the 'Last Tree Is Cut', *IPPR Progressive Review* 24, no. 3 (2017): 176–202.

[28] James O'Connor, *Natural Causes: Essays in Ecological Marxism* (New York: Guilford Press, 1999).

[29] Moore, World Accumulation.

Migrations in the Kinocene

Within our perspective of the Kinocene as a "history" of the Capitalocene that we have previously developed, this section will focus on the approach to a specific mobility: that of human beings, which is contemporarily known as international migrations.

Although simplistic, this last assertion has a very relevant rationale for the field of study we intend to address because mobility and migration are not the same in political terms. Indeed, every migratory process involves mobility, but only contemporary human mobilities that cross international borders between states constitute migrations.[30] Additionally, not every border crossing constitutes, in political terms, a migration, as it can be produced for reasons of tourism, but not even every non-national of a state is considered a migrant, as this category alludes to a subaltern socio-political condition which, in the case of Core-Core or Core-Periphery migration, is not always fulfilled, giving rise to considering these people as mere "foreigners".[31]

Human beings have been mobile for as long as they have existed, which explains their expansion from present-day Ethiopia to their global establishment.[32] Moving from one territory to another is, therefore, an inherent part of our species, whose nomadic character has only mutated into one of more permanent settlement thanks not only to the development of agriculture (made possible by Holocene climatic conditions) but also to the establishment of political regimes aimed at controlling the mobility of their populations.

Departing from an ecocentric framework external to the property relations over land subsequently developed by human societies, human displacement from one territory to another was merely mobility. If no entity claims possession and control of a given territory, then mobility is a mere kinetic exercise. This, however, takes on a different aspect in the world of states, which are political-institutional apparatuses that claim both sovereign control over territories delimited by borders and also the government of their populations to which they grant a membership known as nationality, which distinguishes them from the nationals of other states. As Sayad points out, the "national would only exist in the presence [...] of its opposite or in opposition to its opposite, the non-national"[33]; therefore, the condition of migrant is intrinsically political, as it alludes to the presence of non-nationals in the national territory of the state, a fact that justifies, in the eyes of the state, the application of a whole corpus of specific legislation that limits both their rights and the reasons that authorise their presence in the state.

---

30 Alison Mountz and Shiva S Mohan, 'Human Migration in a New Era of Mobility: Intersectional and Transnational Approaches', *Global Social Challenges Journal* 1, no. 1 (2022): 59–75.

31 Abdelmalek Sayad, *The Suffering of the Immigrant* (Cambridge: Polity Press, 2004).

32 Terry-Ann Jones and Eric Mielants, *Mass Migration in the World-System: Past, Present and Future* (Boulder: Paradigm Publishers, 2011).

33 Abdelmalek Sayad, "Stato, Nazione Ed Immigrazione: L'ordine Nazionale Alla Prova Dell'immigrazione," [State, Nation and Immigration: The National Order Put to the Test by Immigration] in *Abdelmalek Sayad. Per Una Teoria Postcoloniale Delle Migrazioni* [For a Postcolonial Theory of Migration] eds. Gennaro Avallone and Salvo Torre (Catania: Il Carrubo, 2013): 67.

On the one hand, the presence of migrants justifies the state's very existence by giving meaning to the classic functions of territorial and population control attributed to it by Weber.[34] And, on the other hand, by constructing it as a controllable "other", it has the (self-attributed) legitimacy to decide which non-nationals can enter its territory, to carry out which functions and under what conditions.[35] This individualised control of people, which in totalitarian states is widely criticised as an extreme exercise of domination of people's lives, is widely accepted, even unquestioningly, by national societies when it comes to migrants.[36] Thus, democracy, freedom and citizenship become principles that can only be thoroughly enjoyed by nationals, while migrants are controllable subjects who must justify their presence in terms of "utility".[37]

This state pragmatism highly reveals the symbiotic relationship that the state interweaves with capitalism, where both need each other for their survival and smooth functioning.[38] Migrants from the periphery, dispossessed, subalternised and placed at the bottom of the global hierarchy of the world-system, constitute a global reserve army[39] (Bellamy Foster and McChesney, 2018) whose mobility, to a large extent, responds to the accumulation needs of the system.

Thus, if in the Holocene we could witness more or less autonomous mobilities, in the Capitalocene, which began in 1492 with the conquest of America and the global colonisation of the periphery by the core,[40] the great migrations have tended to have a direct relationship with the processes of accumulation. In this sense, we refer to what Jones and Mielants[41] mention as "Mass Migration" as opposed to micro-perspectives that defend migration as a rational decision of individuals.

Throughout the history of the world-system, mass migration has always been related to global processes of accumulation, activating, displacing or promoting (depending on whether the mechanisms employed were direct, indirect or surreptitious) the mobility of large numbers of people either to the colonies or, contemporaneously, mainly from the periphery to the core. Examples of these processes, such

---

[34] Fabio Raimondi, *Migranti e Stato. Saggio Su Abdelmalek Sayad* [Migrants and the State. Essay on Abdelmalek Sayad] (Verona: Ombre Corte, 2016).

[35] Sayad, Suffering of the Immigrant.

[36] Abdelmalek Sayad, "Immigration et 'Pensée d'État" [Immigration and 'State Thought'] *Actes de La Recherche En Sciences Sociales* 129 (1999): 5–14.

[37] Alain Morice "'Choisis, Contrôlés, Placés'—Renouveau de l'utilitarisme Migratoire," ['Chosen, Controlled, Placed'—Revival of Migration Utilitarianism] *Vacarme* 14 (2001): 56–60.

[38] Alex Callinicos, "Does Capitalism Need the State System?," *Cambridge Review of International Affairs* 20,no.4 (2007): 533–49.

[39] John Bellamy Foster and Robert W. McChesney, 'The Global Reserve Army of Labor and the New Imperialism', in *Class: The Anthology*, eds. Stanley Aronowitz and Michael J. Roberts (Oxford: Wiley-Blackwell, 2018): 485–502.

[40] Jason W. Moore, 'Anthropocene, Capitalocene & the Flight from World History: Dialectical Universalism & the Geographies of Class Power in the Capitalist World-Ecology, 1492–2022', *Nordia Geographical Publications* 51 (2022): 123–46.

[41] Jones and Mielants, Mass Migration.

as the Atlantic slave trade or temporary migration programmes in the twentieth and twenty-first centuries, share this logic.[42]

The systemic need for cheap labour is an essential law of capitalism,[43] and people who migrate are directed, thanks to the complicity of states, towards the productive links in the value chains that require their labour force. Thus, the migrant has a reduced margin for manoeuvre due to the limitations generally imposed by his condition. Such restrictions sometimes manifest in not being free to change employers, the labour sector or being unable to leave jobs as this implies being deportable, a mechanism designed to ensure the system's viability, making migrants cheap labour available.

This does not imply that migrants lack agency and are merely a passively available labour force for global value chains.[44] On the contrary, they reveal how people, even under highly oppressive structures, exercise agency, a fundamental dimension for understanding subsequent reconfigurations of the political realities in which they are actors.[45]

Migrations in the Capitalocene are thus torn between their structural activation—the result of the needs outlined by capitalist actors to guarantee the global processes of accumulation—and the life projects that migrants elaborate on. In any case, this tension between the macro and the individual always has the capitalist world-ecology as its point of reference, either because it deliberately activates mobilities to guarantee its perpetuity or because the actors, despite undertaking autonomous migratory processes, often obey or respond to systemic logics.

In a given structure, such as the vertebration of the capitalist order between the core and the periphery, the individual decision to undertake a south-north migration, for example, never occurs outside the system because even the most rationalist perspectives would assume that this decision is motivated by the existence of a global hierarchy and an unequal distribution of wealth. The same is true for those who may be alienated in one territory and decide to exercise upward social mobility in another in response to the capitalist logic to which they are subjected. In short, detaching knowledge of human migrations from the systemic structure in which they occur is an epistemological error.

Dimensions such as job opportunities, the social stability of environments or even the legal frameworks in force are intrinsically related to historical and socio-political factors of the capitalist world-ecology. Hence, in this Kinocenic "age of migrations"[46]

---

[42] Yoan Molinero Gerbeau, 'Mass Migrations across the World-System's History', E-International Relations, 2017, accessed July 17, 2024, https://www.e-ir.info/2017/08/01/mass-migrations-across-the-world-systems-history/.

[43] Jason W. Moore, *Capitalism in the Web of Life: Ecology and the Accumulation of Capital* (New York: Verso, 2015).

[44] Sandro Mezzadra, and Brett Neilson, *Border as Method, or, the Multiplication of Labor* (Durham and London: Duke University Press, 2013).

[45] Oliver Bakewell, 'Some Reflections on Structure and Agency in Migration Theory', *Journal of Ethnic and Migration Studies* 36, no.10 (2010): 1689–1708.

[46] Stephen Castles, Hein De Haas and Mark J. Miller, *The Age of Migration. International Population Movements in the Modern World* (New York: Palgrave Macmillan, 2014).

where processes of accumulation and intensification of mobilities go hand in hand, categorising them around labels such as "economic migration", "forced migration" or "environmental migration" fragments the object of study and impedes its holistic understanding.

In the next section, we will see how migrations, and particularly those constructed as "environmental" or "climate", are the fruit of a cognitive fracturing which, far from being accidental, is aligned with the interests of the state, whose creation of dualisms contributes to maintaining or enhancing its positions in the hierarchy of the capitalist world-ecology.

Breaking the Two Dualisms

Although the climate migrant category has gained increasing recognition within and outside academia,[47] it has been a contentious notion since its inception. Shortly after it was coined in 1985 by Essam El-Hinnawi as an "environmental refugee",[48] the study of displacement caused by environmental degradation *split* into two central positions.[49]

On the one hand, maximalists believed that environmental degradation directly causes large-scale displacement, particularly from the Global periphery to the Global core.[50] Among the works of these maximalists were those by Myers,[51] Jacobson[52] and El-Hinnawi,[53] who estimated that the phenomenon could involve hundreds of millions of people. They regarded environmental degradation as the primary driver of migration and one of the most pressing issues facing humanity in the twenty-first century. These estimates drew attention and gained popularity due to their alarmist nature.

On the other hand, minimalists reject causal determinism, arguing that environmental degradation alone is insufficient to cause displacement. They see migration

[47] E. Piguet, R. Kaenzig, and J. Guélat, 'The Uneven Geography of Research on "Environmental Migration"', *Population and Environment* 39, no. 4 (2018): 357–83, https://doi.org/10.1007/s11111-018-0296-4.

[48] Essam E. Hinnawi and UNEP, *Environmental Refugees. by Essam El-Hinnawi* (UNEP, 1985), https://digitallibrary.un.org/record/121267.

[49] Astri Suhrke, 'Environmental Degradation and Population Flows', *Journal of International Affairs* 47, no. 2 (1994): 473–496.

[50] Suhrke.

[51] Norman Myers, 'Environmental Refugees: A Growing Phenomenon of the 21st Century', *Philosophical Transactions of the Royal Society B: Biological Sciences* 357, no. 1420 (29 April 2002): 609–13, https://doi.org/10.1098/rstb.2001.0953.

[52] J.L. Jacobson, 'Environmental Refugees: A Yardstick of Habitability', *Worldwatch Paper* 86 (1988).

[53] Hinnawi and UNEP, *Environmental Refugees.*

as a complex phenomenon in which the influence of climate is difficult to isolate from social, political and economic factors, which play a predominant role.[54]

Over time, empirical research has shifted the scientific debate towards minimalist viewpoints.[55] Case studies have shown that the type of mobility predicted by maximalists—environmentally determined, transnational, massive and long term—does not align with the observed realities of displacement. Instead, these cases often reveal a more complex and heterogeneous picture. When mobility does occur, which is not always the case, it tends to be primarily seasonal or circular,[56] segmented[57] (affecting specific population groups based on gender or age) and mostly internal.[58] These findings challenge the maximalist thesis that environmental factors are the primary drivers of mobility. On the contrary, they suggest that displacement results from multiple factors.

The difficulty of attributing displacement solely to environmental factors presents both conceptual and methodological challenges.[59] Since climate cannot be identified as the primary driver of displacement—always subordinate to economic and social factors[60,61]—it becomes impossible to distinguish between a conventional economic migrant and a climate migrant.[62]

This conceptual vanishing point has led to a considerable academic effort, with numerous authors attempting to develop a framework to delineate the epistemological

[54] Suhrke, 'Environmental Degradation and Population Flows'.

[55] Andrew Baldwin, Chris Methmann, and Delf Rothe, 'Securitizing "Climate Refugees": The Futurology of Climate-Induced Migration', *Critical Studies on Security* 2, no. 2 (2014): 121–30, https://doi.org/10.1080/21624887.2014.943570.

[56] Xin Lu et al., 'Unveiling Hidden Migration and Mobility Patterns in Climate Stressed Regions: A Longitudinal Study of Six Million Anonymous Mobile Phone Users in Bangladesh', *Global Environmental Change* 38 (1 May 2016): 1–7, https://doi.org/10.1016/j.gloenvcha.2016.02.002.

[57] Michael Mensah, Paul L. G. Vlek, and Benedicta Y. Fosu-Mensah, 'Gender and Climate Change Linkages in the Semi-Arid Region of Ghana', *GeoJournal* 87, no. 1 (1 February 2022): 363–76, https://doi.org/10.1007/s10708-020-10261-w.

[58] T. Xiao et al., 'Complex Climate and Network Effects on Internal Migration in South Africa Revealed by a Network Model', *Population and Environment* 43, no. 3 (2022): 289–318, https://doi.org/10.1007/s11111-021-00392-8.

[59] F. Gemenne, 'Why the Numbers Don't Add up: A Review of Estimates and Predictions of People Displaced by Environmental Changes', *Global Environmental Change* 21 (2011): S41–49, 10.1016/j.gloenvcha.2011.09.005.

[60] S. Mpandeli et al., 'Migration under Climate Change in Southern Africa: A Nexus Planning Perspective', *Sustainability (Switzerland)* 12, no. 11 (2020), https://doi.org/10.3390/su12114722.

[61] Luisa Veronis and Robert McLeman, 'Environmental Influences on African Migration to Canada: Focus Group Findings from Ottawa-Gatineau', *Population and Environment* 36, no. 2 (1 December 2014): 234–51, https://doi.org/10.1007/s11111-014-0214-3.

[62] Suhrke, 'Environmental Degradation and Population Flows'.

foundations.[63,64,65] However, some, like Dun and Gemene, propose avoiding this category. This does not mean they deny the association; on the contrary, they agree that, in the Kinocene, the environment is a cross-cutting issue in life and mobility. Therefore, it is more useful to include climate in migration studies rather than to seek a strict definition of environmental migration.[66] Baldwin, on the other hand, proposes that climate migrants are a futurible, that is, a theoretical postulate without correspondence to a specific migratory phenomenon. Thus, he maintains that the entire narrative on this topic is presented in a conditional future tense.[67]

Criticism of the category addresses a methodological issue due to its lack of conceptual soundness and raises ethical-ideological considerations that question its usefulness. Some authors argue that this phenomenon represents the academic materialisation of the neo-Malthusian nightmares of the global core countries.[68,69] Thus, this line of research would reinforce suspicion of southern populations, instigate border policies and exercise symbolic violence by rendering invisible the political, social and economic reasons behind mobility by attributing these causes solely to climatic or environmental factors.

At the terminological level, it also suffers from an important shortcoming. With the advance of research, constant criticism and the frequent appearance of new conceptual frameworks, a melting pot of terms has been generated to name the phenomenon. Many terms are applied interchangeably without distinguishing between climate, environment or migrant and refugee.

Despite the problems identified, such as theoretical and conceptual fragility and terminological imprecision, these have not been sufficient to exhaust the line of research. On the contrary, as mentioned at the beginning of this section, this line of research is booming and gaining more and more relevance. Despite the knowledge generated on the complexity of the link between mobility and the factors and moderating elements involved, the dominant discourse continues to be aligned with

---

63 Frank Biermann and Ingrid Boas, 'Preparing for a Warmer World: Towards a Global Governance System to Protect Climate Refugees', *Global Environmental Politics* 10, no. 1 (February 2010): 60–88, https://doi.org/10.1162/glep.2010.10.1.60.

64 Dominic R. Kniveton, Christopher D. Smith, and Richard Black, 'Emerging Migration Flows in a Changing Climate in Dryland Africa', *Nature Climate Change* 2, no. 6 (June 2012): 444–47, https://doi.org/10.1038/nclimate1447.

65 F. Florémont, 'Migratory Issues in Climate Adaptation Policies: A New Conceptualisation of Population Displacements?', *Forum for Development Studies* 39, no. 1 (2012): 31–49, https://doi.org/10.1080/08039410.2011.635382.

66 Olivia Dun and François Gemenne, 'Defining "Environmental Migration"', *Forced Migration Review* 31 (1 October 2008): 10–11.

67 Baldwin, Methmann, and Rothe, 'Securitizing "Climate Refugees": The Futurology of Climate-Induced Migration'.

68 Giovanni Bettini, 'Climate Barbarians at the Gate? A Critique of Apocalyptic Narratives on "Climate Refugees"', *Geoforum*, Risky natures, natures of risk, 45 (1 March 2013): 63–72, https://doi.org/10.1016/j.geoforum.2012.09.009.

69 Piguet, Kaenzig, and Guélat, 'The Uneven Geography of Research on "Environmental Migration"'.

maximalist imaginaries.[70] This discourse we refer to as "sophisticated maximalism" recognises the complexity and multidimensionality of the phenomenon but, at the same time, reproduces a reductionist discourse. In this context, reality is discretised by considering independent causes and positioning the environmental element as the main factor above other factors.[71]

We argue that the persistence of the maximalist discourse is due to a series of discursive mechanisms that make the conception of the climate migrant-that is, displacement as an inevitable response to environmental degradation-intuitive and parsimonious. However, this apparent plausibility is not universal; it is framed by specific cultural and historical coordinates reflecting the global core's epistemological foundations. The key mechanisms behind this discursive inertia include two main dichotomies: the human-nature dichotomy (cartesian dualism) and the legitimate mobility-illegitimate mobility dichotomy. The first dichotomy establishes a separation between the human and natural realms, while the second distinguishes between accepted and legitimate forms of mobility versus those considered illegitimate or problematic.

Society-Nature Dualism

The core conflict around climate migrations is the problematic understanding of the nexus between climate and global mobilities. This is directly connected with the epistemological dispute between the Anthropocene and the Capitalocene as a Kinocene. That is, to know the type of agency and structural drivers that sustain environmental change on mobility and ultimately to unravel the linkage between environment and society. This difficulty has resulted in the production of an infinite number of studies aimed at quantifying the explanatory capacity of environmental and climate change as a predictor of mobility.[72,73,74] These attempts are based on a fragmented understanding of reality, separating the environmental and human dimensions. As a result, mobility and environmental degradation are perceived in hegemonic liberal perspectives as independent aspects, maintaining a monotonous and increasing relationship of influence.

---

[70] Bettini, 'Climate Barbarians at the Gate?'.

[71] David J. Kaczan and Jennifer Orgill-Meyer, 'The Impact of Climate Change on Migration: A Synthesis of Recent Empirical Insights', *Climatic Change* 158, no. 3 (1 February 2020): 281–300, https://doi.org/10.1007/s10584-019-02560-0.

[72] K.C. Pradhan and K. Narayanan, 'Does Climatic Risk Induce Labour Migration? Evidence from Semi-Arid Tropics Region of India', *Journal of Public Affairs* 22, no. 1 (2022), https://doi.org/10.1002/pa.2323.

[73] W.M. Wafula et al., 'Factors Influencing Migration and Settlement of Pastoralists in Nairobi City, Kenya', *Pastoralism* 12, no. 1 (2022), https://doi.org/10.1186/s13570-021-00204-6.

[74] K.K. Zander et al., 'Topic Modelling Exposes Disciplinary Divergence in Research on the Nexus Between Human Mobility and the Environment', *Humanities and Social Sciences Communications* 9, no. 1 (2022), https://doi.org/10.13140/RG.2.2.29899.92963.

The truth is that this lack of familiarity with the mobility-nature link is the result of a historical oversight. With rare exceptions, there are few examples throughout the development of migration science in which the environment has been considered as a mechanism of particular relevance for displacement.[75] Although some pioneering authors in the study of migration, such as Ratzel,[76] highlighted the relevance of the environment in human displacement, the main theories developed in the twentieth century have focused mainly on economic, social and political factors, relegating environmental influence on a secondary role.

This blind spot not only exclusively concerns migration studies but extends to all social sciences and dates back to their origins. Based on the negation of the natural sciences, the social sciences have led to obviating or denying natural causes to explain social phenomena. From then on, the process of knowledge construction based on specialisation by disjunctive logic contributed to the fragmentation of knowledge into disciplines, increasing the knowledge gap between the natural sciences and the social sciences.[77]

Like much of the social sciences, migration studies emerged in the nineteenth century under the influence of Enlightenment thought. However, the separation between nature and society has deeper roots, dating back to the sixteenth and seventeenth centuries Scientific Revolution. Jason W. Moore refers to this as "Cartesian dualism"[78] based on René Descartes' separation of mind and body, which established a new framework of thought grounded in dualisms. This dichotomy distinguished society and nature as separate realms governed by different laws and purposes: nature was seen as external, deterministic and devoid of historical agency, while society was understood as moral, rational and progress-oriented. This reconfiguration of these categories implicitly created a hierarchy, positioning society above nature. Cartesian dualism paved the way for capitalist development by framing nature as external, passive and subject to human control, ultimately reducing it to a reservoir of resources available for exploitation in the name of progress.

According to Moore, value production would not only result from accumulation or technical investment to improve productivity in capitalism. Still, it must also constantly establish what he calls Cheap Nature. This includes affordable production and reproduction of food, labour, energy and raw materials. Consequently, capitalism would not simply be an economic or social system but a way of producing nature (human and extra-human) and maintaining a co-production relationship. To achieve

---

[75] Etienne Piguet, 'From "Primitive Migration" to "Climate Refugees": The Curious Fate of the Natural Environment in Migration Studies', *Annals of the Association of American Geographers* 103, no. 1 (2013): 148–162, https://doi.org/10.1080/00045608.2012.696233.

[76] Friedrich Ratzel, *Anthropogeographie*, Bibliothek geographischer Handbücher (Stuttgart: J. Engelhorn, 1882), http://www.archive.org/details/anthropogeograp02ratzgoog.

[77] C. Hamilton, Christophe Bonneuil, and François Gemenne, *The Anthropocene and the Global Environmental Crisis: Rethinking Modernity in a New Epoch*, 2015, https://doi.org/10.4324/9781315743424.

[78] Jason W. Moore, *Capitalism in the Web of Life: Ecology and the Accumulation of Capital* (New York: Verso Books, 2015).

this, it aligns power, production and perception with the production of meaning, a key element in shaping the practices of capital extraction.[79]

The establishment of the Cartesian dichotomy does not imply that the conception of nature has remained unchanged over the past five centuries; on the contrary, it has been continuously redefined through the process of historical meaning-making. The notion of "Cheap Nature" itself led to a critical reflection on the idea of nature as an inexhaustible resource, spurred by scientists from colonial powers in the global periphery who warned of the devastating consequences of unchecked exploitation. This reflection laid the groundwork for what would become the modern environmental discourse.[80]

Scientific discoveries have strongly influenced green thinking.[81] So, the scientific developments of the twentieth century—the birth of disciplines such as ecology and the realisation of climate change, among others—contributed to strengthening green thinking, which began to impact public debate at the end of the century significantly. The formation of green parties in Europe at the end of the 1980s, the successive climate summits in Istanbul and the Rio de Janeiro summit, and the creation of the category of climate migrants are examples.

In the countries of the global core, this sensitivity has gained centrality to the point of becoming one of the main issues on their political agenda. Despite critical versions of contemporary green thinking, such as the metabolic rift, it is still based on solid Cartesian foundations.[82] It is articulated around "green arithmetic" concepts such as ecological footprint, human impact, ecological crisis and sustainability. In this approach, nature is seen as external, passive and inferior, while society is considered the only one with the capacity, action and awareness to save the environment. A paradigmatic example of this Cartesian vision is the prevalence of the Anthropocene concept in academia to describe the contemporary period, which reveals a profound anthropocentrism by placing the world in the orbit of human agency.[83]

The rise of green thought explains the need to proclaim the existence of climate migrations, conceived as one of the most radical implications of climate change. Presenting these migrations as a distinctive phenomenon becomes essential for strengthening the green discourse, as they are positioned as one of its most dramatic impacts on humanity.

Although the concept of "Cheap Nature" has been questioned, the persistence of the Cartesian dichotomy in how we understand nature limits our comprehension of its true role in people's lives and, consequently, the link between mobility and

---

[79] Moore.

[80] Richard H. Grove, 'Colonial Expansion, Tropical Island Edens and the Origins of Environmentalism, 1600–1860, Cambridge and New York' (Cambridge University Press, 1995), 1600–1860.

[81] Yoan Molinero Gerbeau, *El medioambiente en las Relaciones Internacionales* [The Environment in International Relations] (Síntesis (Madrid, España), 2022).

[82] Moore, *Capitalism in the Web of Life.*

[83] Álvaro San Román and Yoan Molinero-Gerbeau, 'Anthropocene, Capitalocene or Westernocene? On the Ideological Foundations of the Current Climate Crisis', *Capitalism Nature Socialism* 34 (2023): 39–57, https://doi.org/10.1080/10455752.2023.2189131.

environmental change. This Cartesian divide fits perfectly with the dominant push–pull model in migration studies, as it allows for the fragmentation of reality into independent factors, where nature becomes just another variable in econometric models. Once this association is established, it is proclaimed as the defining feature of this type of mobility.

The world-ecology paradigm and the overcoming of the Cartesian dichotomy imply recognising that the influence of what is called nature is transversal to all social mobilities and phenomena. The boundary between society and nature is a fiction, a simplification of reality in the process of enlightened rationality. Instead of seeking to validate the category of climate migrations by including a factor in a model, it is more relevant to develop a situated knowledge of the synergies between human activity and the environment in which it takes place to understand the Kinocene feature of the Capitalocene.

Dualism of Legitimate and Illegitimate Mobility

By analogy, Andrew Baldwin, Christiane Fröhlich and Delf Rothe's assertion that mobility is a crucial term for understanding the Anthropocene[84] could be extrapolated to the concept of the Capitalocene. In this context, the traffic of ideas, goods and people reaches unprecedented levels in terms of distance and speed, resulting from the expansion and consolidation of capitalism as the generator of a particular type of nature (human and extra-human).

Mobility is fundamental to the market economy, as trade itself emerged at the intersection of cultures and trade routes. Throughout history, numerous economic powers have thrived by becoming hubs of intense mobility, such as major commercial ports and cities like Venice in the thirteenth to seventeenth centuries, Amsterdam in the seventeenth century and London in the nineteenth century. Mobility has always been a sine qua non for the functioning of capitalism. The expansion into new markets, pursuing investment opportunities and acquiring cheap natures—raw materials, energy, food and labour—have been crucial for capital accumulation throughout history. In the context of contemporary global capitalism, this dynamic is even more pronounced, as it depends on transportation and telecommunications to sustain financial and production networks on a global scale, enabling the expansion and effective functioning of interconnected economies.

However, although many have characterised globalisation as a period defined by freedom[85] of movement and openness,[86] since the end of the last century, it

[84] Andrew Baldwin, Christiane Fröhlich, and Delf Rothe, 'From Climate Migration to Anthropocene Mobilities: Shifting the Debate', 2019, https://doi.org/10.1080/17450101.2019.1620510.

[85] Liam O'Dowd, James Anderson, and Thomas M. Wilson, eds., *New Borders for a Changing Europe: Cross-Border Cooperation and Governance* (London: Routledge, 2003), https://doi.org/10.4324/9780203508435.

[86] Ronen Shamir, 'Without Borders? Notes on Globalization as a Mobility Regime', *Sociological Theory* 23, no. 2 (June 2005): 197–217, https://doi.org/10.1111/j.0735-2751.2005.00250.x.

has been observed that the contemporary mobility paradigm is based on mobility and immobility. Anthony Elliott proposes the idea of "regimes of immobility",[87] suggesting that mobility coexists with stillness. This immobility can be obligatory, such as the length of stay required to obtain a residence permit for immigrants, or voluntary, such as the rest of the privileged.

Many studies on mobility focus on how people's ability to move, or motility, varies significantly, revealing a significant asymmetry.[88,89] Variables such as social class, gender and nationality are key determinants that explain differences in access to spaces. Jørgen Ole Bærenholdt introduces the term Governmobility to describe a mode of governance based on the control of mobility, which employs a wide range of explicit strategies, such as borders, controls and documents.[90]

This set of controls and coercive measures on mobility is central to what Shamir calls the mobility regime, designed to regulate mobilities that he considers suspicious.[91] Glick Schiller extends Shamir's idea, suggesting that the mobility regime is not a single system but several, which are dynamic and change over time.[92] States, with liberal democracies as a prominent example, have developed these regimes primarily through the distinction between citizens and undocumented aliens. Thus, two types of people are established: those with a high capacity for movement and those whose mobility is seen as potentially dangerous or subversive[93] according to State interests to position themselves in the higher ranks of the world-ecology.

States are not the only actors in regulating and legitimising mobility regimes. Cultural processes, such as the discourses and attitudes that produce processes of inclusion or exclusion, also play an important role in this area. These processes contribute to the construction of the mobility regime, influencing the identity's subjectification: the definition of a person and where he or she is allowed to be.

By adopting the concept of the Capitalocene to understand the contemporary planetary period and using the world-ecology as a frame of reference, mobility (human and extra-human) can be considered a fundamental dimension within the ecosystem, essential for the co-production of nature, i.e. for shaping the world as we know it. In this context, the Capitalocene implies a particular mobility regime regulated by the guiding principle of capital accumulation. This generates a dual scheme of mobility

87 Anthony Elliott, 'From Mobilities to Mobile Lives and beyond: The World According to John Urry', in *Mobilities and Complexities* (Routledge, 2018).

88 Mimi Sheller, 'Uneven Mobility Futures: A Foucauldian Approach', in *Traces of a Mobile Field* (Routledge, 2017).

89 Malene Rudolf Lindberg et al., 'Uneven Mobilities and Epistemic Injustice: Towards Reflexive Mobilities Research', *Mobilities* 19, no. 1 (2 January 2024): 134–50, https://doi.org/10.1080/17450101.2023.2244682.

90 Jørgen Ole Bærenholdt, 'Governmobility: The Powers of Mobility', *Mobilities* 8, no. 1 (1 February 2013): 20–34, https://doi.org/10.1080/17450101.2012.747754.

91 Shamir, 'Without Borders?'.

92 Nina Glick Schiller and Noel B. Salazar, 'Regimes of Mobility Across the Globe', *Journal of Ethnic and Migration Studies* 39, no. 2 (1 February 2013): 183–200, https://doi.org/10.1080/1369183X.2013.723253.

93 Shamir, 'Without Borders?'.

(legitimate/suspected), in which high mobility is encouraged from the global core to activate markets and create new opportunities for exploitation. In contrast, the mobility of other natures (human and extra-human) is restricted to ensure access to cheap nature (raw materials, food, labour and energy at low cost).

This guiding principle, which determines the distinction between legitimate and illegitimate mobility, has great explanatory power. It allows us to understand the creation of dualities at the macro, meso and micro levels, since capitalist culture permeates society at all these levels. Moreover, it offers a broader understanding of mobilities that the traditional approach to mobility regimes does not contemplate. For example, Cohen et al. refer to "subversive mobilities" to point out that there are forms of mobility that escape the hegemony of mobility regimes as traditionally defined.[94] These mobilities, such as the illegal trafficking of people and goods, are carried out in defiance of state strategies of control. However, from the perspective of the world-ecology mobility regime, these activities merely respond to the principle of capital accumulation: they seek cheap labour (human trafficking) and raw materials (drug trafficking) to maximise profits. Although these mobilities defy states, their success is based on their ability to generate profits.

As for climate migrants, this duality contributes to mobility from the global south being seen as suspect. One of the main critical arguments against this category is that it reproduces a stereotypical image of the subjects: impoverished, racialised and belonging to the global periphery. Piguet argues that scientific production on climate migrants follows a North–South logic, where the North studies the South. When thinking about climate migrants, it is difficult to conceive of a subject from the global core in that category. This is because the internalised duality that legitimises mobility is linked to capital maximisation. Although people from the global core are also affected by the environmental crisis, their mobility is not perceived as suspicious, even though they contribute significantly to accelerated socio-ecological degradation.

In short, mobility (human and extra-human) is a key element for the establishment and development of capitalism, therefore the Kino-centric view of the Kinocene proposed by Nail is key to a proper reading of the Capitalocene.

Conclusion

This chapter discusses the category of climate migrants as a knowledge device specific to the structures of thought of the capitalist world-ecology. Where power, production and perception converge to legitimise a particular regime of mobility articulated on maximising capital accumulation. Such an analysis has led to a discussion of the different frameworks for understanding contemporary mobility, an analysis of structural factors and their relationship to mobility within the world-ecology, and

[94] Erik Cohen, Scott A. Cohen, and Xiang (Robert) Li, 'Subversive Mobilities', *Applied Mobilities* 2, no. 2 (3 July 2017): 115–33, https://doi.org/10.1080/23800127.2017.1305217.

the exposure of the symbolic structures on which the notion of the climatic migrant is founded.

As a result of the growing concern with climate change, a melting pot of new concepts has emerged within the social sciences to define the current period of environmental transformation, replicating Crutzen's formula, i.e. adding the suffix "cene" to the idea from which they conceive life to be articulated. The proliferation of new categories is limitless, fuelled by the profitability of the promise of citations, which means that many of them are hastily coined without sufficient theoretical underpinning.

Among the proposals discussed is the notion of the Kinocene, through which Tomas Nail tries to propose a non-anthropocentric scheme of understanding based on the dynamic essence of substance.[95] According to our analysis, the idea suffers from certain limitations in terms of its sustainability and understanding the conjuncture of climate change within the capitalist system. We consider that there are currently only two "cenes" with sufficient rigour and explanatory capacity: the Anthropocene and the Capitalocene. Between these, we favour the notion of the Capitalocene, as it has more scope to explain structural dynamics driving social phenomena such as human mobility. However, we consider the Kinocene sub-narrative, or history, to contribute novel ideas and new perspectives to the broader and more robust theory of the Capitalocene.

In particular, we rescue two ideas from Nail's thinking. The first is that the essence of the world is dynamic. Therefore, the world-ecology presupposes a specific mobility regime whose regulation is oriented towards maximising the accumulation of capital.

The Capitalocene mobility regime is sustained by both material and symbolic structures—such as the border, passport, nationality, currency—which include the category of climatic migrant and the symbolic foundations on which it rests. These foundations are two dualisms: legitimate-illegitimate mobility and nature-society.

On the one hand, the nature-society duality, which originated at the beginning of modernity with the development of rationalist philosophy, imposed a specific way of understanding society and nature as independent, albeit hierarchically organised in spheres. In this conception, nature was relegated to a reservoir of raw materials destined for the development of the capitalist system. On the other hand, the dualism of legitimate and illegitimate mobility establishes different levels of autonomy and freedom of movement depending on the origin of the individual. This framework for understanding mobility makes it possible to establish a favourable regime for capitalist development, in which the high mobility of capital and capitalists contrasts with the restricted mobility of labour. Encouraging population mobility in the global core is fundamental to capitalism's material and symbolic expansion and capital accumulation. Limiting mobility in the global periphery is vital to maintaining a cheap labour force and natural resources by controlling the reserve army and perpetuating the economic asymmetry that allows for low commodity prices.

[95] Nail, Kinocene: 375.

Consequently, the concept of the climate migrant is a thinking device that not only builds on two fundamental symbolic structures of contemporary world-ecology, but also re-legitimises and reinforces them, adding problematic nuances. These nuances include an apocalyptic potential for the global periphery's mobility and the invisibility of the underlying political and economic issues surrounding its mobility.

In the following, we propose two strategies aimed at overcoming this dualism that contribute to the perpetuation of the world-ecology and, consequently, promote the contemporary socio-ecological crisis:

1. The first strategy consists of disarticulating the logical schemes that sustain the category, and the rhetoric constructed around the notion of climate migrants, such as the dualisms: nature-society and legitimate-illegitimate mobility.
2. The second strategy is introducing the concept of CoPe (Core-Periphery) Mobility, to break the duality between legitimate and illegitimate mobility. Cope mobility would be an antagonistic concept to migration. For the moment, we do not propose a closed definition but rather an intuition of what could be a new line of research that contributes to questioning the mobility of the global core, especially that which, directly or indirectly, perpetuates the world-ecology and contributes to the impoverishment and dependence of the countries of the global periphery. This category would, therefore, encompass a broad set of different forms of mobility, such as binge tourism, sex tourism or so-called digital nomads.

Chapter 6
From the Anthropocene to the Techno-Westernocene

Álvaro San Román Gómez

The "Anthropocene" is not an innocent and neutral scientific characterization of a geological period, but rather the original title of a meta-narrative that relates, organizes, and comprehensively explains a diversity of events, from rising temperatures to human gene editing. In this narrative, the concepts of Nature and human being, the anthropological characterization of humanity, undergo a profound and dangerous transformation due to the axiological centrality of technology. In order to draw attention to the fundamentally ideological character that drives the anthropocenic discourse as a "species narrative" or "speciesist narrative", that is, as a narrative that essentializes contingent characters of the human species, as the technological progress, we will confront it with the Capitalocene narrative understood as a "cultural narrative", as a narrative that emphasizes the cultural character of all human adaptation to the environment. Going one step further, we will see how the Capitalocene narrative itself makes two important anthropocenic concessions. The first of these is to consider technology from a double essentialist-utilitarian perspective, that is, as a reality that specifically defines and differentiates human beings, and simultaneously, as a tool that derives its value from the use that is made of it. The second concession, aligned with the previous one, will be to assume the post-natural characterization of Nature, with its concept of hybridization, which will ultimately justify, as a new mode of the old hybris, the total subjugation of holocene Nature to the designs of technology. Finally, we will try to highlight the debt that these proposals have with profoundly dualist premises and with the Western project of dominating Nature through the technologization of the environment, being able to affirm that the problem is not the Anthropocene, not even the Capitalocene, but the Techno-Westernocene. We will conclude by showing how the theses presented here have clear implications for global politics.

Á. San Román Gómez (✉)
Universidad Nacional de Educación a Distancia, Madrid, Spain
e-mail: alvarosanroman@fsof.uned.es

M. Robson (ed.), *The Anthropocene, Ontopolitics, and International Relations*,
https://doi.org/10.1007/978-981-96-2925-1_6

Society-Nature Interrelationship

As is well known, in the early 2000s the Nobel laureate in chemistry Paul Crutzen coined the term "Anthropocene".[1] With this concept, the chemist came to synthesize the powerful idea that humanity had acquired such power as to become a geomorphic force. Beyond glaciations, volcanisms, or neptunisms, humanity would today be positioned as the main cause of planetary modification. Indeed, there seems to be a broad consensus underpinning the "Anthropocene" concept that the degradation of the biosphere, the collapse of biodiversity and biogeochemical alterations are the responsibility of the anthropos. However, far from such a consensus and against this "*speciesist narrative*", the "*cultural narrative*" emerges as an alternative. Each geological narrative will establish its own foundational moment as an interpretative proposal for our period. Each of them will assume certain beliefs about what humans, Nature and their relationship to each other are, thus creating the possibility of justifying the narrative. But of course, what the great majority of stories take as the great lesson learned is the essential interrelationship between society and Nature.

For some years now, the influence of climate change on human lives has become indisputable. The alteration of the planet is evidence that we are leaving the Holocene behind. When the emissions produced during the lifetime of an average American cause the death or serious suffering of two people in the world[2]; when three square metres of Arctic ice melt for every trans-oceanic flight[3]; when the normal development of children's cognitive capacities suffers due to exposure to pollution in big cities,[4] then we have reached what Peter Sloterdijk called "the end of cosmic carelessness".[5] We have reached that moment in which human beings can no longer evade the awareness of their power over and dependence on this planet they inhabit. A moment in which it becomes fundamental both to be concerned about the discordant relations that human beings maintain with the environment, and to be occupied in reflecting on and responding to the reasons for this change of period.

In this reflexive moment, the intertwining of society and Nature emerges with full evidence precisely when society is assaulted by certain natural threats associated with climate change, a product of the actions of that same society. It is this dialectical relationship that the Capitalocene theorist Jason Moore, a Marxist ecologist, highlights through his proposal of the concept of *Oikeios*. For Moore, *oikeios*, etymologically

---

[1] Paul Crutzen, *The 'Anthropocene'*. Earth System Science in the Anthropocene, (Berlin: Springer, 2006), 13–18.

[2] John Nolt, "How Harmful Are the Average American's Greenhouse Gas Emissions?," *Ethics, Policy & Environment* 14, No. 1 (2011): 3–10.

[3] Notz, Dirk and Stroeve, Julienne. "Observed Arctic Sea-Ice Loss Directly Follows Anthropogenic CO_2 Emission," *Science* 354 (2016): 747–750.

[4] Catherine Derom, Evert Thiery, and Steven Weyers, "Residential Green Space and Child Intelligence and Behavior Across Urban, Suburban, and Rural Areas in Belgium: A Longitudinal Birth Cohort Study of Twins," *PLOS Medicine* 17, No. 8 (2020).

[5] Peter Sloterdijk, ¿Qué sucedió en el Siglo XX? [What happened in the Twentieth Century?] (Madrid: Siruela, 2018), 17.

"favourable place", "is a way of naming the creative, historical, and dialectical relation between, and also always within, human and extra-human natures ".[6] *Oikeios* can be better understood as "Nature-as-matrix", i.e. as the continuum of Nature-society as relational totality. From this perspective, we would move away from what Moore calls the "green arithmetic" used by certain ecologist leftists, whose formula would be "Society + Nature", understanding both as autonomous and separate entities, and we would insert ourselves into a new paradigm whose formula would rather be "Society = Nature", understanding their relationship as a dynamic unity. This perspective is also shared by anthropocenictheorists such as the Spanish political scientist Manuel Arias Maldonado, who attests to how "environmental factors come to be considered an active subject with social impact". Thus, history would not only be made by human beings, "but also by resources such as water or coal".[7] In the twilight of cosmic carelessness, at this interstitial moment between two periods, the hybrid co-evolution of Nature and society would be confirmed to the detriment of the dualist human/Nature opposition, which would result in an alternative conception of the latter, in a post-nature in which there would be no place for any hint of purism or naturalistic romanticism. Nature in the Anthropocene will cease to be seen as the scene of human evolution, not even as the home of our species, nor as the sum of forests, seas, and mountains, and will become an absolute and abstract constancy that would pulsate equally in the concrete of factories, in wind farms or in the engines of our cars. Thus, it will turn out that "Nature can neither be saved nor destroyed, only transformed",[8] so all that remains for us to do is to embrace this *socio-natural mix* and abandon the former idea of a pre-human Nature, thus immersing ourselves in the *Oikeios* as a certain post-nature. From here we would be prepared to assume anthropogenic influence as inevitable, as being the blind dynamism of Nature influencing itself, and it would only remain to discuss how this inevitability is embedded in the fate of the planet.

Capitalocene: The Cultural Narrative

The era of climate emergency delivers us as an eternal truth the interdependence of the human and its environment. "All life makes environments. All environments make life ".[9] This means that niche creation is inevitable. However, if we look at the ways in which nature is transformed and organized to create a suitable environment for the development of social life, we can see that there is a wide variety of ways. Indeed, if

[6] Jason Moore, *Capitalism in the Web of Life: Ecology and the Accumulation of Capital* (New York: Verso, 2015), 53.

[7] Manuel Arias Maldonado, Antropoceno, la política en la era humana [*Anthropocene, politics in the human era]* (Madrid: Taurus, 2018), 30.

[8] Jason Moore, *Capitalism in the Web of Life: Ecology and the Accumulation of Capital* (New York: Verso, 2015), 53.

[9] Moore, *Capitalism in the Web of Life: Ecology and the Accumulation of Capital*, 65.

"thus food-getting and familymaking (...) they are ways of environment-making ",[10] we could consider with Moore that the very *Bourse* of Amsterdam in 1602, or Wall Street in the twenty-first century, would in itself be "a way of organizing Nature".[11] If hunter-gatherer societies had their own ecology understood as "the set of diverse relationships between species and environments",[12] capitalism, as a very specific way of managing resources, ordering and exploiting Nature, species and individuals on a global scale, would be a world-ecology that has as its outstanding characteristic an altered climate that puts at risk the very survival of all societies, whether they are capitalist or not. Capitalism, therefore, would not be a social system with its own ecological regime, but would *itself* be an ecological regime, for its *raison d'être*, accumulation, would not merely be a socio-economic process with environmental consequences, but rather "a way of bundling human and extra-human natures".[13]

According to Jason Moore, this regime had its founding moment in 1492. It consecrated an epochal change that led the world from feudalism to capitalism through its characteristic *modus operandi*, the colonization of space. Because, in effect, it was geographical expansion that made possible the transition to capitalism. The discovery and subsequent colonization of the American continent by Europeans brought relief from the periodic ecological crises suffered by feudalism, which had to respond to a continuous demographic increase with dwindling tracts of productive land. Thus, the expansion of the areas of influence of the European powers through the exercise of power and the stimulus provided by the accumulation of natural resources established a new global and globalizing order defined by the *frontier movement*.[14] Through the conquest of new frontiers, which Moore divides into horizontal (absorbing continents) and vertical (mining), "vast reserves of Nature's gifts were enclosed, appropriated, and put to work in the global circuit of capital".[15]

But capitalism, in this vicious circle from the desire for accumulation, through resource crises, to colonialist expansionism, has a terminal structural problem of its own, which is that "capital's demand for cheap nature tends to increase faster than its capacity to secure it",[16] and this because "the main mercantile frontiers have been terminated".[17] Under the capitalist regime, humans have reached the ends of the earth and have capitalistically organized all resources, all means of production and all labour forces, bringing us to climatic collapse. Therefore, it is not so much the human species but capitalism itself as a "differentiated civilization",[18] which is

---

[10] Moore, *Capitalism in the Web of Life: Ecology and the Accumulation of Capital*, 62.

[11] Moore, *Capitalism in the Web of Life: Ecology and the Accumulation of Capital*, 100.

[12] Jason Moore, La trama de la vida en los umbrales del Capitaloceno [*The web of life on the threshold of the Capitalocene]*. (Mexico: Bajo Tierra, 2020), 235.

[13] Jason Moore, *Capitalism in the Web of Life: Ecology and the Accumulation of Capital* (New York: Verso, 2015), 56.

[14] Moore, *Capitalism in the Web of Life: Ecology and the Accumulation of Capital*, 346.

[15] Jason Moore. *La trama de la vida*, 173.

[16] Moore, *The web of Life on the Threshold of the Capitalocene*, 237.

[17] Moore, *The web of Life on the Threshold of the Capitalocene*, 252.

[18] Moore, *The web of Life on the Threshold of the Capitalocene*, 169.

mainly responsible for the climatic disaster, the protagonist of this new geohistorical period. So much so that we could conclude with Moore that "the problem is not the Anthropocene, but the Capitalocene".[19]

Yet, the historical and cultural evidence does not seem enough for those who obstinate in making prevail what we have called the species narrative. The theorists of the Anthropocene, executing in the intellectual order the colonialist exercise that capitalism practices in the material order, phagocytize the conceptual universe of the Capitalocene itself, including it in their narrative as a mere epigraph of the "*Human Era*". According to Arias Maldonado, "if we combine the history of the species with that of its social and economic institutions, we may find that capitalism seems much more 'natural' than its quietist alternatives". We will discover that "certain traits of the species *predispose* it to organise itself in a capitalist way".[20] This premise of capitalist naturalness is reiterated in political science and post-naturalist or even post-theological philosophy. With his characteristic cynicism, the British philosopher Timothy Morton also emphasizes this idea of the capitalist inherence in the human soul. Against theoretical critics who underline the origin of climate change as Western, more precisely American, Morton argues that, insofar as American capitalism lives on consumer desire, and desire is itself a human *predisposition*, the problem of the Anthropocene boils down to the innocence of our desiring structure. "Although the desire for it emerged in America first, chronologically, it turns out that everyone wants air-conditioning".[21] It would seem, therefore, that any critic of the Anthropocene would be tragically argumentatively entangled with Time and its becoming, by making a synchronic critique of a problem that is ultimately diachronic, if the fundamental issue of the Anthropocene is the human being itself. Indeed, if it turned out that time jumps were possible, even Neanderthals, according to Morton, would discover themselves capitalists; "Neanderthals would have loved Coca-Cola Zero".[22]

Anthropocene: The Speciesist Narrative

Without detracting from the truths that the theoretical framework of the Capitalocene may highlight, the Anthropocene can impose its narrative by repeatedly falling into the fallacy of generalization. "If 'capitalism', reason their theorists, 'in its accumulative and expansionist zeal has brought us this far, but it is human beings who *make* capitalism, then', they conclude, 'it must be humanity that is responsible'". In a game of misleading abstractions, the Anthropocene considers humanity the

---

[19] Moore, *The web of Life on the Threshold of the Capitalocene*, 234.

[20] Manuel Arias Maldonado, *Anthropocene, politics in the human era* (Madrid: Taurus, 2018), 60.

[21] Timothy Morton, "How I Learned to Stop Worrying and Love the Term Anthropocene." *Cambridge Journal of Postcolonial Literary Inquiry* 1, No. 2 (2014): 4 https://doi.org/10.1017/pli.2014.15.

[22] Morton, "How I Learned to Stop Worrying and Love the Term Anthropocene," 4.

sum of cultural and subjective individualities. This is why, instead of situating the foundational moment of our period in 1492, it can situate it in the Neolithic revolution, with the transition from hunter-gatherer nomadism to sedentary agriculturalism,[23] or even in the evolutionary milestone of the mastery of fire as narrated by the Pyrocene variant, for which "the firestick morphed into an Archimedean lever by which to move the planet".[24] Thus, under anthropogenic premises, there would be a continuum between the discovery of fire and the new fires burning in oil wells, from primitive man cracking stones in his cave to modern man drilling oilfields. Indeed, there seems to be a subtle thread woven through history that has inevitably brought us global warming and the sixth mass extinction. That thread would be Human Nature itself, thought from the theoretical framework of the Anthropocene. This proposal is not so much historical or cultural as anthropological. To explain the current order, anthropocenics resort to postulating an "alternative anthropology" that tends to naturalize the techno-capitalist cultural aspects.

For many of the Anthropocene narrators, it is often impossible to conceive of the human species inhabiting the planet without transforming it, thereby harming other living beings and their environment. From their anthropological point of view, the domination and self-interested modification of nature is presented as an innocent adaptive necessity. Thus, the critical development of the Capitalocene narrative would be nothing more than the detailing of what would ultimately be the Anthropocene, because "the optics of the species emphasises the universal drive towards aggressive adaptation",[25] coincidentally with very capitalist traits, through a long and natural process of colonization and transformation.

Therefore, it would be incongruous, according to this narrative, to accuse humanity of breaking its ecological limits since the *frontier movement* would not simply be the way of life of capitalism but would be a purely human movement. A movement of aggressive adaptive expansion, but always and above all through technology, since, in fact, to this human Nature would also belong, as a last and fundamental anthropological characteristic, "our technological vocation",[26] so that the humanization of the environment would pass directly through its technological modification to adapt it to human needs. In other words, while clownfish lives in anemones and chimpanzees among branches, human beings would live among cars, computers, and smartphones. Our home automation, or Apple's headquarters in Silicon Valley, would not be qualitatively different from the Yanomami´s *shabono*. They would be simply regional and local variations of a universal impulse that is they are other ways of transforming the natural environment along a progressive historical line. The only difference for anthropocenics is that while the Yanomami have a simple or

[23] Ruddiman, William, "The Anthropogenic Greenhouse Era Began Thousands of Years Ago," *Climatic Change* 61 (2003): 261–293. https://doi.org/10.1023/B:CLIM.0000004577.17928.fa.

[24] Pyne, Stephen J. "The Planet is Burning." November 20, 2019. https://aeon.co/essays/the-planet-is-burning-around-us-is-it-time-to-declare-the-pyrocene.

[25] Manuel Arias Maldonado, *Anthropocene, Politics in the Human Era* (Madrid: Taurus, 2018), 21.

[26] Antonio Diéguez, Transhumanism: The technological quest for human enhancement, (Barcelona: Herder, 2017), 147.

pre-modern type of technology that allows them to maintain a relationship with the environment that gives them a certain *sense* of unity and harmony, we moderns have developed our technology by complexifying the environment to the point of making our relationship with Nature complex. Since the differences between the Yanomami people and Silicon Valley workers are reduced to a mere amount of technologization of the environment according to the intensity of their aggressive adaptation to it, it is fair to assume that, over time, these differences will disappear as societies converge in the unstoppable and inevitable process of globalization. A species that adapts by expanding its complex technology cannot but tend to globalize, to exert its dominance over the environment, species, and individuals.

In short, our aggressive adaptation, driven by our expansionist zeal, leads inevitably to the globalization of the technological exploitation of species and environments, to the techno-colonization of spaces and other human groups, to global techno-capitalism, and "it is a mistake, therefore, to denounce the human colonization of the planet as if it were something alien to Nature itself when it is nothing less than a process by which Nature works on itself through human culture".[27] The speciesist narrative has, as we see, the virtue of justifying the excesses of techno-capitalist culture; firstly, by diluting its responsibility in the vast agential mesh that post-nature represents, since we would only be vehicles of an encompassing Nature that would manifest itself in multiple forms, in our twenty-first-century human case, in a techno-capitalist way. At the same time, it attributes all responsibility directly to human individuals as a whole, in a perverse reverse psychology whereby anyone who denounces the evils of this system can be labelled a misanthrope.

In this process of "essentialisation" of catastrophe, we see a perverse dialectical movement being replicated once again, and that is that our survival seems to be endangered by how our species survives. The Anthropocene, the anthropogenization of the planet, seems to set the conditions of possibility for the disappearance of these Anthropos. However, there is a tendency to find a way out of this *cul de sac* by way, once again, of our "technological condition".

The Anthropocene paradigm draws on the sources of techno-utopianism, of technological programme, which assumes that every human need and problem is satisfied and solved technologically. Thus, paradoxically, it will turn out that only more Anthropocene will solve the problems of the Anthropocene. The solution will be to embrace the Anthropocene, immersing ourselves in technological innovation applied to natural systems and even to human beings themselves.

As Sloterdijk would say, "technology has not yet said its last word",[28] and it will come to preach it in the realm of Nature and of the human being himself.

[27] Manuel Arias Maldonado, Anthropocene, politics in the human era (Madrid: Taurus, 2018), 139.

[28] Peter Sloterdijk, *What happened in the Twntieth Century?* (Madrid: Siruela. 2018), 28.

Anthropocene and Posthuman Tech-Nature

Today, technological advances such as those envisaged, for example, in the field of geoengineering of the climate,[29] are such that one tends to think hopefully of the possibility of direct climate manipulation so that climate change itself could, in turn, be changed through the management of solar radiation and the reduction of carbon dioxide in the atmosphere. The Anthropocene narrators, starting with Crutzen himself, are convinced that we can turn the tide of the Anthropocene "by vastly improved technology and management, wise use of Earth's resources".[30] Ecological restoration seems possible, always, and only if we take into narrative that "there are enormous technological opportunities".[31]

Given this panorama, it is therefore sustainable to think that our current system of production and exploitation can be reconciled with the scarcity of available natural resources, since we can modify them at the whim of our immediate needs. Certain apologists of the Anthropocene, such as Crutzen himself, wonder, in fact, whether the transformative capacity of human beings themselves can manage to increase the degree of substitutability of natural capital, in a conjecture similar to that made by transhumanist theorists such as Max More, for whom "intelligent use of biotechnology and nanotechnology and the opening of new frontiers in space, can remove resource constraints and discharge environmental pressures".[32] From this technutopian perspective, it would therefore seem that planet Earth, and the resources we need from it, are not suffering the terrible crisis that environmental theorists have been denouncing for decades. Nature may have its own limits, its own boundaries, but just as the limit imposed by an amputated leg can be overcome through the implantation of a prosthesis, so Nature could overcome its limits with biotechnological implants. What would transgenic food, or even the colonization of Mars or the Moon, be if not the technological thrust of Nature itself? According to Peter Sloterdijk, "as long as the Earth and its biosphere are regarded as an unincrementable singularity, modern civilization's exploitative behaviour of expression and comfort must appear as an unforgivable irrationality".[33] Thus, for anthropocenic technutopian, the problem for environmentalist critics would ultimately consist in not knowing what the Earth's body can do if we set it to dance to the tune of what our technology can do. Again, in a perverse dialectic, we see that we, natural beings, generate artificial technological realities that we apply to Nature to begin a kind of co-evolution, inaugurating at last a new way of contemplating and understanding our relationship with the natural environment and its species, a monistic post-natural, or hybrid way of relating to

---

[29] José Luis Granados Mateo, "Technoscientific futures: new challenges around solar geoengineering," *Revista CTS* 16 (2021): 11–40.

[30] Paul Crutzen, *The 'Anthropocene'*. Earth System Science in the Anthropocene, (Berlin: Springer, 2006), 17.

[31] Crutzen, *The 'Anthropocene'*. Earth System Science in the Anthropocene, 17.

[32] Max More, "The Extropian Principles, Version 3.0" Extropy Institute, 1998. https://www.mrob.com/pub/religion/extro_prin.html

[33] Peter Sloterdijk, *What happened in the Twntieth Century?* (Madrid: Siruela. 2018), 28.

Nature. Finally, the paradigm of hybridization emerges as the paradigm that replaces the dualistic and outdated romantic vision of a supposed "natural purity". Today, as geologist Jan Zalasiewicz states, "the Anthropocene represents a new phase in the history of both humankind and of the Earth, when natural forces and human forces became intertwined, so that the fate of one determines the fate of the other".[34] Indeed, in an evolutionary symbiotic embrace, and still thanks to technology, it can now be said that "Nature is us".[35] Theorists such as the Timothy Morton, and even Jason Moore with his concept of *Oikeios*, are now postulating such embraces. With their proposals they would help us to interpret the Anthropocene, the hybridization that beats in its breast, the place that Nature and we occupy in it, telling us about the decentring of human agency within a vaster network of techno-living relations. On this horizon, technology would then not be a human prerogative, a mere contingent concretion of a cultural modality, but an innocent mode by which nature transforms itself in its many facets, from the human to the vegetable. This is why Morton states that "it is not sensible to say that 'technological' genetic manipulation is bad, as if the breeding of stallions were not technical manipulation. Breeding is a form of technology. Fields and ditches are technology. Monkeys that catch termites with a stick are technological".[36] Morton comes clean, and writes: "I am sorry to say it, but he have lost that soft, squishy, irrational, authoritative Nature. Yes, we have lost it, because it never even existed".[37] Now, at the dawn of the Anthropocene, it turns out that Nature is not the soft lines of those hills that taught Albert Camus more than he ever learned from science[38]; Nature is no longer that which taught Faustus to meet his brothers in the quiet forest, in the air and in the water[39]. Now and always, Nature also turns out to be the petrol stations decorating the motorways, the gigantic windmills, or the ocean liners breaking the sea horizon. Our relationship with Nature, which technology came to complicate by eliminating the sense of unity that prevailed before, is once again simplified by the grace of technological hybridization. Like evolved Yanomamis, the humans of the anthropocenic narrative regain their harmony with the environment, suddenly remembering that everything is connected, and not just through roots, sap, ocean currents, or rivers of lava, but through fibre-optic cables carpeting the ocean floors, oil pipelines, and satellites orbiting the planet. The Anthropocene would thus mean a certain revival of *new*

[34] Jan Zalasiewicz, Mark Williams, Will Steffen, and Paul Crutzen. "The new world of the Anthropocene," *Environmental Science & Technology 44*, (2010): 2231 https://doi.org/10.1021/es903118j.

[35] Paul Crutzen and Schwägerl Christian. 2011. "Living in the Anthropocene: Toward a New Global Ethos." *Yale Environment 360* (http://tinyurl.com/kcbjrgm).

[36] Timothy Morton, *El pensamiento ecológico* (Barcelona: Paidós. 2018), 113.

[37] Morton, *El pensamiento ecológico*, 167.

[38] Albert Camus, *The Myth of Sisyphus* (Buenos Aires: Editorial Losada, 2004), 33—"...this science that was supposed to teach me everything ends in the hypothesis (...). The soft lines of these hills and the hand of the twilight on this agitated heart teach me much more."

[39] Johann Wolfgang von Goethe, *Faust*. Part 1, Scene 14, "Forest and Cavern" (Madrid: Aguilar, 1968). 133. "I gaze, as in the bosom of a friend. The ranks of living creatures thou dost lead Before me, teaching me to know my brothers in air and water and the silent wood."

age mysticism, riding on siliceous chips and 5G, which learns to see its brothers in the Kirobo robot, in Alexa, In Siri, in the cloned kitten. Today, the ecological landscape "is a vast *mesh* of interconnections without a centre or set contours"[40], a *mesh* populated by old acquaintances and new inhabitants, by "*strange strangers*",[41] from wild animals, to cloned pets, kitchen robots, and artificially intelligent robots. In the Anthropocene, agency is intertwined in a myriad of entities and processes, in which attribution of responsibility is meaningless; agency, adds Jason Moore, "is, rather, an emergent property of definite configurations of human activity with the rest of life. And vice versa ",[42] and therefore "deep reflection on collaboration is a must for ecological thinking. All strange strangers are already collaborating. In the city where I live", Morton comments, "there are thousands of crows using cars and streets as nutcrackers".[43]

From the above perspective, the Anthropocene is intended to be the period in which Nature ceases to be an ontological otherness, and begins to be a totalizing selfhood, overcoming its capitalist ontological limitations through technological overpowering, thanks to which a hybrid environment will emerge in which ancient, cloned and robotic species will collaborate in the unity of a globalized system, in a supposed monistic reality, which, as we see, will give theoretical justification to posthuman positions.

Anthropocene and Transhumanism

So far, we have seen that, for leading anthropocene theorists, combating climate change can be achieved through technological manipulation of the climate, just as its effects on humans can in turn be reversed by genetically manipulating humans themselves to adapt them to an increasingly hostile environment.[44] This is the transhumanist project, and this is where it connects with anthropocenic proposals. For a technological world, a technological human; for a hybrid Nature, a hybrid humanity. And if the good thing about the technological modification of climate change itself is that "what remains to be done can be accomplished without major changes in patterns of production and consumption",[45] the good thing about the transhumanist project, as the philosopher José Antonio Diéguez critically observes "that one can remain a normalized member of society and devote one's leisure time carefully to personal

[40] Timothy Morton, *The Ecological Thourght* (Barcelona: Paidós, 2018), 25.

[41] Timothy Morton, *The Ecological Thourght* (Barcelona: Paidós, 2018).

[42] Jason Moore, *Capitalism in the Web of Life: Ecology and the Accumulation of Capital* (New York: Verso. 2015), 55.

[43] Timothy Morton, *The Ecological Thourght* (Barcelona: Paidós. 2018), 130.

[44] S. Matthew Liao, Anders Sandberg and Rebecca Roache, "Human Engineering and Climate Change," *Ethics, Policy and Environment* 15, No. 2 (2012): 206–221.

[45] Michael Keary, "The New Prometheans. Technological Optimism in Climate Change Mitigation Modelling," *Environmental Values* 25, No. 1 (2016): 8. https://doi.org/10.3197/096327115X14497 392134801.

cultivation",[46] i.e. it would be enough to agree. After all, it seems to be a feature of our species to modify and colonize new frontiers; and what is the human being if not a frontier, a vast field of modification? Alongside the horizontal and vertical frontiers through which our techno-capitalism expands, in the form of colonization (thanks to maritime, land or air transport), and in the form of exploitation and extraction of natural resources (thanks to drilling rigs or mining), a third frontier emerges, the ontological frontier, which allows the modification of the being of things thanks to biotechnological tools. It would confirm to us that Nature that surrounds us has no *purity* whatsoever, that Nature in us has no impassable limits. "We have decided that it is time to amend the human constitution"[47] stated the transhumanist pioneer Max More decades ago, and fifteen years later Crutzen and Schwägerl take up the same discourse, stating that "we are taking control of Nature's realm, from climate to DNA".[48] More himself, in a premonitory way, confirmed that "over the coming decades we will pursue a series of changes to our own constitution, initiated with the tools of biotechnology".[49] And he was not wrong. Anthropocenic humanity is today taking the difficult art of human evolution into its own hands. In a tremendous technological paroxysm, today, "one might say that technology is the next biology".[50]

The great discovery that opened up this *crack* in the nature of the given through which the manufactured by technology slips in, and which made clear the modifiable nature of human reality, was made in 2012, when the research team led by the biochemist Jennifer Doudna developed the genetic tool known as CRISPR-Cas9. This disruptive discovery, "the genetic scalpel", associated with the results of the sequencing of the human genome in 2003, realized the oldest ambitions of human manipulation. "Now, for the first time ever, we possess the ability to edit not only the DNA of every living human but also the DNA of future generations-in essence, to direct the evolution of our own species".[51] Today, mind and body pretend to be dissociated and modified at the whim of our desires, and in response to our needs.

In this sense, in the words of Max More, the anthropocenic transhumanists "advocate using science to accelerate our move from human to a transhuman or posthuman condition",[52] a condition that finally leaves behind the undesirable aspects of what until now was considered too human, such as ignorance, death, and ageing, but also

---

[46] Antonio Diéguez, *Transhumanism: The Technological Quest for Human Enhancement* (Barcelona: Herder, 2017), 21.

[47] Max More, *A Letter to Mother Nature. The Transhumanist Reader*. (Oxford: Wiley, 2013), 449. https://doi.org/10.1002/9781118555927.ch41.

[48] Paul Crutzen and Schwägerl Christian, "Living in the Anthropocene: Toward a New Global Ethos," *Yale Environment 360* (2011). http://tinyurl.com/kcbjrgm.

[49] Max More, *A Letter to Mother Nature. The Transhumanist Reader* (Oxford: Wiley, 2013), 449. https://doi.org/10.1002/9781118555927.ch41.

[50] PETER HAFF, "Technology as a Geological Phenomenon: Implications for Human Well-Being," *London: Geological Society, Special Publications* (2014). https://doi.org/10.1144/sp395.4.

[51] Jennifer A. Doudna and Samuel H. Sternberg, *A Crack in Creation: Gene Editing and the Unthinkable Power to Control Evolution* (New York: Houghton Mifflin Harcourt, 2017), 10.

[52] Max More, "The Extropian Principles, Version 3.0" Extropy Institute, 1998. https://www.mrob.com/pub/religion/extro_prin.html.

obsolescence with respect to the machines that our technological vocation brings us, or our inability to live in adverse and changing climates. Indeed, intellectual, physical, and genetic "improvement" "will be necessary if we are to compete with super-intelligent machines",[53] as M. Liao, A. Sandberg, and R. Roach conclude "biomedical modification of humans" will be the solution "so that they can mitigate and/or adapt to climate change".[54]

As we can see, the anthropocenic panorama is clearly sustained by transhumanist, i.e. technologicist, assumptions. The Anthropocene, therefore, would delineate itself as the result of a planetary modification practised by the human species, in the exercise of its natural traits of aggressive adaptation by technological means, organized in a capitalist way. Unfortunately, this modification has generated enormous devastation in the biosphere, but fortunately it is intended to be reversed through the over-powering of Nature by means of technologization, not only of the environment, but also of the human being himself, who is emerging as a hybrid, transhuman subject.

Native Paradigm and Paradigm of Use

The Anthropocene would not have achieved its ideological hegemony if it had not been for a deeply rooted evaluative tendency in our Western globalized society, which creates the conditions for the possibility of elaborating posthuman and transhumanist proposals such as those described above: Today there is a tendency to consider technology from a double essentialist-utilitarian perspective. On the one hand, technology, and its concrete historical development, is assumed to be that reality which specifically defines and differentiates human beings, and on the other it is placed beyond good and evil, in a neutral zone with no capacity to generate other values than those of the end which justifies its use. Therefore, nuclear energy, fossil fuels, AI, the whole industry that supports their production, as well as the kind of life they sustain, would not be intrinsically evil, nor would they possess any anti-human traits with alienating effects. Technological artefacts, from houses to cars, would be the way in which human beings adapt to the environment. Technology seems to be rabidly human, which is why it would be foolish to accuse it of being the main cause of environmental degradation. Here, essentialism justifies utilitarianism.

Thus, a comprehensive critique of the technological system today is confronted with two difficulties. The governance of what we would call "the native paradigm", according to which subjects assume that technology is the human environment, the native land that the human species always carries with it, giving meaning and purpose to its existence, in such a way that renouncing the computer or the smartphone would

[53] Antonio Diéguez, *Transhumanism: The Technological Quest for Human Enhancement* (Barcelona: Herder, 2017), 132.

[54] S. Matthew Liao, Anders Sandberg and Rebecca Roache, "Human Engineering and Climate Change," *Ethics, Policy and Environment* 15, No. 2 (2012): 206–221.

result in the renunciation of its own humanity. And, on the other hand, the rule of what we would call "the paradigm of use", according to which the evil derived from the techno-scientific apparatus cannot be imputed in essence to the apparatus itself but to the use that is made of it.

It is into the trap of this latter paradigm, which hides the substantive character of technology as a project of domination of nature, that the Capitalocene narrative would also fall, preventing it from a more radical diagnosis of our geohistorical moment. But it could not be otherwise, judging by Karl Marx's own perspective on the matter. Assuming the "paradigm of use", Marx will come to affirm that the contradictions and antagonisms inseparable from the capitalist application of machinery do not arise out of machinery as such, but out of its capitalist application.[55] It is from this approach that Moore writes that "while nineteenth-century industrialization surely accelerated the degradation of Nature, this line of reasoning, in its narrow conception, attributes undue weight to technological progress".[56] "The Industrial Revolution is important", he concedes, "but the central problem is imperialism",[57] so he concludes that "it was capitalism, not industrialization alone, that was the real culprit".[58]

However, contrary to this paradigm, we will argue that the problem is to think that technology is not the problem. If, indeed, it was to turn out that technology is not the problem, but that the problem derives from the way it is used, we might think that it would be enough to change the predatory capitalist use of the technological apparatus to redress the disasters caused by it. Unfortunately, however, if we look at the historical development of the alternative to capitalism par excellence, the Soviet alternative, we will see that its geohistorical heritage includes disasters that are just as devastating as those caused by the capitalist system. Thus, the desertification of the Aral Sea, as well as the infamous Chernobyl nuclear disaster. What Capitalism and Soviet Communism would therefore fundamentally have in common would be a technologistic view of existence in which technology is the mode and means to be free, either to produce and consume commodities or to stop making them. Both alternatives assume that technology is the answer to human problems, rather than understanding it as the major human problem to be solved; both alternatives assume the paradigm of use and the native paradigm.[59]

And yet, contrary to the "native paradigm", the fact that people today arrive in a technologized world does not make that world as natural as water is for fish. The Spanish philosopher José Ortega y Gasset, in his meditations on technology, explains that, since coming into the world, the human being finds himself "surrounded by a

---

[55] Karl Marx, *Capital A Critique of Political Economy Volume One* (Great Britain: Penguin Books, 1982), 569.

[56] Jason Moore, *The Web of Life on the Threshold of the Capitalocene* (México: Bajo Tierra, 2020), 45.

[57] Moore, *The Web of Life on the Threshold of the Capitalocene*, 271.

[58] Moore, *The Web of Life on the Threshold of the Capitalocene*, 45.

[59] The thesis according to which Soviet communism was a state capitalism does not invalidate the thesis here sustained on the sharing of a technologistic ontological thesis at the theoretical level in both Marxism and liberalism, although the historical embodiment of both systems has been substantially different.

fabulous quantity of objects and procedures created by technology, which form a first artificial landscape so dense that it hides the primary Nature behind it, he will tend to believe that, as it is, everything is there for its own sake: that the automobile and the aspirin are not things to be made, but things, like the stone or the plant, which are given to man without any effort on his part",[60] he will therefore tend to think that the technological environment is essentially the human environment. This valuational tendency is deeply rooted in our worldview because of our Western technological fetishism, which forces us to perceive "tangible objects as given and separate from the invisible networks of power relations and domination in which they are embedded".[61] But the historical pre-existence of our technological world does not imply its ontological or axiological pre-eminence.

From the Paradigm of Hybridization to Technologicist Paradigm as Techno-Western Wolrdview

The force-idea that operates as the soul that drives the machine, from the founding myth of Prometheus, that shapes the paradigm of the use, and the native paradigm, the worldview that animates a world that, in the end, is self-deprecating at techno-bites, is that technology is the anthropological fatum and therefore an absolute value. Its ontological and axiological pre-eminence mutually imply each other. Indeed, it is paradoxically the axiological primacy of technology that allows it to be valued solely in terms of its use, and to breathe life into the "paradigm of use". The underlying emotive argument could be summarized as follows: Our technology is so valuable that it must necessarily be absolutely innocent. Its claimed lack of intrinsic value, its ontological innocence, is what makes it so valuable. Technology is good for everything, but nothing can be blamed on it. And in turn, this innocence is endorsed by the principle that technology is human destiny, to the extent that "there is no man without technology"[62].

Thus, there is a circle of argument, which goes from the paradigm of use to the native, from utilitarianism to essentialism, in which, no matter where we enter, we find ourselves deeply entangled. As human beings live, breathe, and become persons among technological artefacts, it will turn out that the human biotope will be technology; in turn, technology is the human habitat, because it is what human beings do to survive; being the mode by which humans survive, it therefore has the character of an anthropological universal condition, that is, it is human destiny. Thus, if technology is the human environment, and at the same time that which human beings do in order to live, that is, it possesses a transcendental anthropological character, none of the technological artefacts, nor the system they configure, can be susceptible

[60] José Ortega y Gasset, *Meditation Technique* (Madrid: Revista de Occidente, 1964), 95.

[61] Hornborg, Alf, "*The Political Ecology of the Technocene*," The Anthropocene and the Global Environmental Crisis (London: Routledge 2015), 61. https://doi.org/10.4324/9781315743424-5.

[62] José Ortega y Gasset, *Meditation Technique* (Madrid: Revista de Occidente, 1964), 39.

to moral evaluation. Technology, as an environment and anthropological universal, would always be beyond good and evil. Indeed, if technology is an anthropological condition, it would not be intrinsically bad or good, but a condition from which we can live better or worse, just as language is not something bad or good but the condition for saying good or bad things.

It is in this vicious argumentative circle that the paradigm of hybridization lives, stands, and prospers. Because only if technology is the product of the innocent evolutionary process of Nature, hypostatized by the human being, can it be considered innocent. And, finally, only from here, can the programme of the absolute technologization of the environment, of posthuman ecological techno-restoration, of transhuman techno-adaptation to the environment, be interpreted as the consummation of universal monism, under the paradigm of hybridization.

Because it is not precisely the monistic idea of a pre-existing harmony, but the simpler and more savage dualistic idea rooted in the West that existentially splits humanity from Nature and vice versa, that underlies the technologistic programme of the Anthropocene. In fact, the trap set by any discourse akin to hybridization is to pass off dualism as monism. But hybridization is never monistic, precisely because if hybridization ecologically implies anything, it is a basic dualism, that is, two primordial entities that later merge to form a single entity.[63] The hybridization to which the narratives we expose thus far appeal masks the operational reality of a worldview that remains and needs to remain profoundly dualistic, profoundly Western. Nature and Technology are indeed unified, but under the rule of technology,[64] and technology is always dualistic as the ultimate expression of an agonistic and antagonistic relationship between humans and their environment. Technology is "the energetic reaction against nature", it is always "the reformation of nature",[65] and not the summum of the historical unfolding of monism.

As the German physicist Werner Heisenberg explains, "from the very start we are involved in the argument between nature and man" and technoscience will not be a manifestation of the suture between the two, but precisely a manifestation of this dualism.[66] This being so, Crutzen and Shwägerl's assertion "it's no longer us against 'Nature'. It's we who decide what nature is what it will be",[67] is ontologically a tautology. That humans, in the Anthropocene, decide what Nature is, simply implies a change *in* the power relations that humans have had for centuries with their environment, in no case is it a change *of* such relationships. Thus we can understand hybridization as a new mode of the old *hybris*, the new way in which

[63] And this is not to mention the ontological difficulty of considering technology as a type of organism, if one wishes to speak properly of "hybridisation," for indeed, in an ecological or biological sense, hybridisation is only a crossing between two organisms.

[64] Jeremy Baskin, "Paradigm Dressed as Epoch: The Ideology of the Anthropocene," *Environmental Values* 24, No. 1 (2015): 9–29. https://doi.org/10.3197/096327115X14183182353746.

[65] José Ortega y Gasset, *Meditation Technique* (Madrid: Revista de Occidente, 1964), 28.

[66] Werner Heisenber, *The Physicist's Conception of Nature* (London: Hutchinson Scientific and Technical, 1958), 24.

[67] Paul Crutzen and Schwägerl Christian. 2011. "Living in the Anthropocene: Toward a New Global Ethos." *Yale Environment 360* (http://tinyurl.com/kcbjrgm).

we can continue imposing our terms on Nature. Hybridization is not, therefore, the hypostasis of monism, but of the most antediluvian lack of moderation, of that Greek hubris that one day made the human species believe it could transgress limits, and above all, alter the relationships established prior to its emergence. It is no coincidence that technology, as a project for the domination of Nature, being mythologically and historically associated with human arrogance since Prometheus gave fire to humanity, finally finds its culmination in the effective technological alteration and modification of natural relations under the paradigm of hybridization. That we are the ones who decide what Nature will be, means that, definitely, nature no longer imposes anything on us, and that the imposition is now technological. And technology, Heisenberg continues, "intervenes radically in the relationship of nature to man, radically changing his environment and thus bringing him face to face with the scientific aspect of the world",[68] leading him to submit, as we have seen, to the "native paradigm", and to devise systems of thought and projects that essentialize technology anthropologically and ecologically, until a neo-environment[69] is born, or more specifically a techno-environment in which Nature is subjugated, *ergo* dominated, by technology.

Under the weight of a genuine and necessary concern about the sinister paths that have taken our geohistory, posthumanism, in his attempt to think of the human being beyond the Western tradition, has made the serious mistake of incorporating in its analysis the native paradigm and the paradigm of technological use, instead of including them as a fundamental part of said tradition, and in that measure its concept of hybridization can be considered a debtor of Western dualism, and therefore, of the old *hybris.* The very Donna Haraway, in her famous Ciborg Manifesto, insert herself comfortably in this paradigm, and thus entangles in the confusion between the hybridization of the Ciborg and the overcoming of dualism. Technology, he says, "have made thoroughly ambiguous the difference between natural and artificial".[70] But this difference is only ambiguous if it is accepted in advance that technology can assume characteristics belonging to nature itself, that is, only if technology can be understood as our native land. And given that ambiguity, to overcome it, Haraway chose to incorporate into his thesis the paradigms of technological use and native. Instead of seeing technological hybridization with animals and machines as the definitive consecration of the technological domination project of Western dualism, Haraway chose to see in hybridization the opposition to that project. In the same way that, according to the Anthropocene narrative, only through technology can we get out of the ecological dead end, for Haraway only the "cyborg imagery can suggest a way out of the maze of dualisms"[71] in which we are deeply lost. "Perhaps",

[68] Werner Heisenber, *The Physicist's Conception of Nature* (London: Hutchinson Scientific and Technical, 1958), 19.

[69] Agostino Cera, "The Technocene or Technology as (Neo)Environment," *Techné: Research in Philosophy and Technology* 21, Nos. 2–3 (2017): 243–280. https://doi.org/10.5840/techne201710472.

[70] Donna Haraway, *A Cyborg* Manifesto (University of Minnesota Press, 2016), 11.

[71] Haraway, *A Cyborg* Manifesto, 67.

Haraway continues, we can learn from our fusions with animals and machines how not to be Man, the embodiment of Western logos"[72] as if the embodiment of this Western logos were only the man, and not his creation, the technological apparatus.

For Haraway, as for those who hold positions close to posthumanism like Moore himself, technology is not the problem, but, as we have already seen, capitalism to the extent that it replicates and encourages Cartesian dualism. And yet, the Cartesian assumption of an unlimitedly tolerant external Nature, which, according to Sloterdijk, "gave the cosmic carelessness of human beings after the Industrial Revolution a longer life",[73] could have its foundational moment, no longer "in the age of Columbus",[74] but in the cradles of Western civilization, which is why it has been possible to affirm that the problem is not the Anthropocene, but the Westernocene.[75] This would have its mental inflection in ancient Greece, when the mechanistic image of Nature and the idea of technology as a method of domination began to take shape,[76] and would later settle with Christianity, when the biblical adage "fill the earth and subdue it" (Genesis 1:28) was materially established through effective techno-capitalist exploitation.

Therefore, we agree with Heisenberg that it is the West as a project, and capitalism as western, and not the human species, nor any other culture, "has gained immeasurable power", fundamentally "by applying the Western idea of controlling and exploiting natural resources" by means of technoscience.[77] The technologicist paradigm, as a meta-paradigm that subsumes the paradigms of use, native, and hybridization, is the worldview that precedes and enables the capitalist exploitation of Nature, the technological modification of the environment, and the biogenetic manipulation of human beings in the context of climate change. On this basis we should radically re-qualify the way we conceptualize our geohistorical period, and affirm, going beyond the Capitalocene, that the problem is not the Anthropocene, not even the Technocene,[78] the problem is ultimately the Techno-Westernocene, the profound identity established between technological development as a means of domination of Nature, and the very history of what was, is, and will be, Western culture.

---

[72] Haraway, *A Cyborg* Manifesto, 52.

[73] Peter Sloterdijk, *What Happened in the Twntieth Century?* (Madrid: Siruela, 2018), 19.

[74] Jason Moore, *The Web Of life on the Threshold of the Capitalocene.* (México: Bajo Tierra, 2020), 271.

[75] Álvaro San Román and Yoan Molinero-Gerbeau. "Anthropocene, Capitalocene or Westernocene? On the Ideological Foundations of the Current Climate Crisis." *Capitalism Nature Socialism* 34, No. 4 (2023): 39–57. https://doi.org/10.1080/10455752.2023.2189131.

[76] Pierre Hadot. 2021. *The Veil of Isis. Essay on the History of the Idea of Nature* (Barcelona: Alpha Decay, 2021).

[77] Werner Heisenber, *The Physicist's Conception of Nature* (London: Hutchinson Scientific and Technical, 1958), 66.

[78] Hornborg, Alf, "*The Political Ecology of the Technocene*," The Anthropocene and the Global Environmental Crisis (London: Routledge 2015), 61. https://doi.org/10.4324/9781315743424-5.

Conclusion

The Anthropocene is first and foremost a narrative, a meta-narrative, a vision of the world. It is a project that works with theoretical assumptions that are difficult to verify, due to ontological and anthropological prejudices, which simultaneously have as their basis and objective the idea and the project of essentializing both the project of infinite capitalist accumulation and Western technology: thinking of it as an anthropological universal, instead of thinking of it as a cultural concretization of certain human abilities and desires.

On the other hand, assuming post-natural positions, the project of hybridization, possibly implies the noble task of taking charge of the place of human beings in nature, but, affirming that "Nature cannot be saved or destroyed, only transformed"[79]; unfortunately, holocenical natural concretions, such as the baiji dolphin or the Pyrenean glaciers, are abandoned to the misfortune of their effective destruction and disappearance because they share the same ontological and theoretical space with this naturally hyper-technologized human being. Therefore, diluting the natural and cultural boundaries, that is, natural and technological, ultimately implies naturalizing the Techno-Western disaster, and lending credibility to the Anthropocene narrative.

We believe that we must understand technology for what it is; far from being the sum of technological devices and the factories that light them, it is a worldview, a hegemonic cultural system, the Techno-Western system.

Against the Anthropocene, to assume the narrative of the Techno-Westernocene is to assume that technology is the problem, that technology is what effectively separates us from nature, and not what allows us to move harmoniously towards it. That is why we cannot move towards a true post-natural symbiosis riding on the technological vehicle. We cannot immerse ourselves in the dark indifference of an all-encompassing nature if we always carry with us the blinding light of the technological flame. If technology is the evidence of an effective ontological separation between human beings and Nature, it will never be the way to connect with it.

The triumph of the Anthropocene narrative is to pass off as Nature that which is its destruction. When Moore speaks of "co-production with Nature",[80] when Sloterdijk claims that on this path our world "could become a hybrid planet where more is possible than conservative geologists believe",[81] when Nature ceases to be trees, mountains, rivers, and seas, and expands to include factories and airports, brain chips and bionic arms, then the climate crisis reaches its zenith and Nature is definitively ruined. If we do not feel that with the hybridization of this techno-Nature something or of enormous importance is being lost, then all is lost. Nature is in danger when we stop thinking about it and surrender its concept and care to the narrative of the Anthropocene.

---

[79] Jason Moore, *Capitalism in the Web of Life: Ecology and the Accumulation of Capital* (New York: Verso. 2015), 53.

[80] Jason Moore, *The Web of Life on the Threshold of the Capitalocene* (México: Bajo Tierra. 2020), 235.

[81] Peter Sloterdijk, *What Happened in the Twntieth Century?* (Madrid: Siruela. 2018), 28.

The consequences of its effective ideological operationalization, through the techno-movement frontier, are already visible on the international stage:

In 1967, in an outburst of anthropocentric humility in the history of international relations, the " The Outer Space Treaty"[82] was signed. It clarified that the Moon and other celestial bodies could not be the object of national appropriation or sovereignty claims. In 2030, the Moon will finally be exploited and colonized territory, and the asteroids will be the scene of the new American space mining according to its Asteroid Act[83].

In 2013, Facebook, now Meta, techno-colonized Myanmar, "bringing it" into the twenty-first century by putting in place the necessary infrastructure to connect the country to the internet for free, on the condition that it would be the main search engine, information manager and operator in the country. In 2017, the Rohingya genocide was shown to be directly related to Meta's strategy[84].

In 2012 the genetic tool CRISPR was created. In 2018, Chinese scientist He Jiankui created the first genetically edited children, and there is still no international agreement on its legislation.

In 2024, the company Neuralink performed the first brain implant in humans. Advances in neurotechnology, in association with the development of AI, have led to the emergence of categories such as "neuro-rights".

In 2025, the AI war between the American and Chinese versions has begun, leading us into a new era where technologies surround us, in which finally Tyranni "leave the body free and sets to work directly on the soul".[85] At present, the geoengineering referred to above, in its main methods, carbon dioxide removal (CDR) and solar radiation management (SRM), poses a clear challenge within the framework of International Relations,[86] by implying the need for "a global regime of stabilization" of Spaceship Earth,[87] which in turn implies, at the dawn of our posthuman era, the hypocritical exercise of setting up the human being, once again, as the representative of what the Cosmos, as a union of the human and the non-human, desires and needs.[88] Finally, AI is the undoubted "game changer" at the dawn of

---

[82] https://treaties.unoda.org/t/outer_space.

[83] https://uscode.house.gov/view.xhtml?req=granuleid:USC-prelim-title51-section51303&num=0&edition=prelim#:~:text=A%20United%20States%20citizen%20engaged,obtained%20in%20accordance%20with%20applicable.

[84] https://www.amnesty.org/es/latest/news/2022/09/myanmar-facebooks-systems-promoted-violence-against-rohingya-meta-owes-reparations-new-report/.

[85] Max Horkheimer, Theodore Adorno, *Dialectic of Enlightenment, Philosophical Fragments* (California: Stanford University, 2002). 105.

[86] Corry, O and Kornbech, N (2021) "Geoengineering: A New Arena of International Politics," In: Chandler, D, Müller, F and Rothe, D, (eds.) International Relations in the Anthropocene: New Agendas, New Agencies and New Approaches. Palgrave Macmillan, Switzerland, pp. 95–122. https://doi.org/10.1007/978-3-030-53014-3.

[87] Peter Sloterdijk, *What Happened in the Twentieth Century* (Madrid: Siruela. 2018),18.

[88] Erika Cudworth And Stephen Hobden, "Complexity, ecologism, and posthuman Politics," *Review of International Studies*, 39, (2013): 643–664 British International Studies Association https://doi.org/10.1017/S0260210512000290.657.

the Techno-Westernocenee, posing challenges in all spheres of existence, especially with regard to the environment, insofar as its material sustainability requires an ever-increasing consumption of energy, and therefore an ever-increasing exploitation of natural resources.

These are the most important concretizations of what will ultimately turn out to be the Techno-Westernocene, a geological period driven by a historical moment in which the Techno-West with its techno-environmental techno-imposition, can be thought of as Horkheimer and Adorno thought of the Enlightenment, being totalitarian as only a system can be,[89] to the extent that technology as a system aspires to totality.

[89] Max Horkheimer, Theodore Adorno, *Dialectic of Enlightenment, Philosophical Fragments* (California: Stanford University, 2002). 18.

Chapter 7
Beyond the Capitalocene

Matthew Robson

Introduction

For António Guterres, the current Secretary-General of the United Nations, 'climate change is the mother of all stealth taxes paid by everyday people and vulnerable countries and communities.'[1] Guterres is surely right, and amidst this deeply troubling socio-ecological context, vigorous debate continues unabated over how best to conceptualize our current historical epoch.

Originating from the natural sciences, the concept 'Anthropocene' has now assumed a ubiquitous prominence throughout academic discourse, even making its way into environmental policy frameworks.[2] The 'Anthropocene' points to the central role that human beings have had as a 'physical force' in altering the Earth System and in the process bringing about a new geological era.[3] Such a diagnosis appears to dissolve the ontological/epistemological binary divisions separating 'culture' from 'nature,' with all the political, economic, and ethical implications which this entails. Nevertheless, consensus on the validity of this epochal designation has been hard

[1] A. Guterres (2024) 'Secretary-General's special address on climate action "A Moment of Truth",' *United Nations Secretary-General*, 5th June 2024.

[2] United Nations Development Program (2020) *The Next Frontier—Human Development and the Anthropocene*, New York: UNDP.

[3] P. J. Crutzen and E. F. Stoermer (2000) 'The Anthropocene,' *Global Change Newsletter*, 41 (May 2000), pp. 17–18.

M. Robson (✉)
Faculty of Political Science and Public Administration, Chiang Mai University, Chiang Mai, Thailand
e-mail: matthew.r@cmu.ac.th

M. Robson (ed.), *The Anthropocene, Ontopolitics, and International Relations*,
https://doi.org/10.1007/978-981-96-2925-1_7

to come by, with authors preferring a host of alternative neologisms: e.g. 'Capitalocene,'[4] 'Chthulucene,'[5] 'Kinocene,'[6] 'Technocene,'[7] or the 'Androcene.'[8] In bringing global capitalism back to the centre of analyses the emergence of the concept 'Capitalocene' has been a particularly welcome development. Through a 'World-ecology' framework, authors such as Jason Moore point to the 'environment-making' of global capitalism and the pernicious binaries such as society-nature which facilitate the appropriation and exploitation of non-human nature.[9]

Whilst work on the 'Capitalocene' has certainly been useful in illustrating the role that global capitalism plays in altering the Earth System (e.g. climatic and environmental change), this chapter addresses two perceived limitations in how the concept has featured in this literature. The first of these concerns the under-theorization of global capitalism-state relations. This is a significant shortcoming if we acknowledge the crucial role states play in enabling capitalism and its exploitation of non-human nature.[10] A second and related issue concerns how transformation of current state-capitalism relations might occur, and the role that human beings could play in these processes. If, as implied in World-ecology frameworks, any resort to the agential capacities of human beings would succumb to 'Cartesian dualism' (i.e. the now untenable separation of 'culture' from 'nature'), then the human potential for contesting and transforming state-capitalism relations into more benign forms appears to be curtailed.[11] This is ultimately a question of ontopolitics since the ontological assumptions underpinning notions of a 'Capitalocene' seem to hamper the political prospects for any human-driven transformative change.

This chapter explores answers to these two issues by turning to the assemblage work of Gilles Deleuze and Felix Guattari. The aim shall not be to simply rehearse the geo-philosophical conceptual architecture of these thinkers or to seek definitive answers. It is the more modest task of putting some of their concepts to work in search of novel insights into these crucial, yet under-explored, facets of how the global capitalist economic system is implicated in climate and environmental change. Deleuze and Guattari's original oeuvre will need to be supplemented to bring it up to

---

[4] J. W. Moore (2015) *Capitalism in the Web of Life: Ecology and the Accumulation of Capital*, New York: Verso. See also Molinero-Gerbeau and Pérez this volume.

[5] D. Haraway (2016) *Staying with the Trouble: Making Kin in the Chthulucene*, Duke University Press.

[6] T. Nail (2019) 'Forum 1: Migrant Climate in the Kinocene,' *Mobilities*, 14(3), pp. 375–380.

[7] A. Hornborg (2015) 'The Political Ecology of the Technocene: Uncovering Ecologically Unequal Exchange in the World-System,' in C. Hamilton, F. Gemenne, & C. Bonneuil (Eds.) *The Anthropocene and the Global Environmental Crisis*, Routledge. See also San Román this volume.

[8] J. Vuillerod (2021) 'The Anthropocene is an Androcene: Three Ecofeminist Narratives,' *Nouvelles Questions Féministes*, 40(2), pp. 18–34.

[9] Moore, 'Capitalism in the Web of Life'.

[10] Although see J. W. Moore (2011) 'Ecology, Capital, and the Nature of our Times: Accumulation & Crisis in the Capitalist World-Ecology,' *American Sociological Association*, 17(1), pp. 107–146.

[11] K. Jansen and J. Jongerden (2021) 'The Capitalocene Response to the Anthropocene,' in H. Akram-Lodhi, K. Dietz, B. Engels, and B. McKay (eds) *The Edward Elgar Handbook of Critical Agrarian Studies*, Cheltenham: Edward Elgar, p. 643.

date. I will do so firstly by taking a de-tour through some groundbreaking Marxist IR scholarship, and secondly through engagement with recent work which has similarly made use of Deleuzoguattarian frameworks.

The Capitalocene and Its Discontents

The emergence of the concept 'Anthropocene' has shaken up many academic fields, including that of International Relations (IR). Paul J. Crutzen made this designation of our current geological epoch in the 'Geology of mankind,' accompanied by an account of how human beings comprise an 'environmental force' for the ways in which they have been disrupting the Earth System.[12] Working within posthuman and complexity frameworks, authors have also recognized the agential capacities of the Earth systems, something which renders the culture-nature binary structuring knowledge in fields, such as IR acutely problematic.[13] Human beings, it would seem, are inextricably bound up with their natural environment. The Anthropocene's implications for IR are such that the transgression of its disciplinary limits becomes seemingly inevitable, as authors pursue alternative explanatory frameworks and potential sources of resistance to the existential crisis.[14]

While there are few sceptics remaining who deny the deleterious impact that human beings are having in bringing about climatic and environmental change, the adequacy of the concept 'Anthropocene' to explain these phenomena is another matter entirely. From a political perspective, at the very least, the abstract and imprecise agent denoted by 'Anthropos,' or 'human made' climate change, is troublesome and susceptible to instrumentalization. Authors working from within Marxist and/or Feminist theoretical traditions have argued, logically, that it is insufficient to ascribe the responsibility for global heating to 'human beings' per se. Some sections of humanity have undoubtedly had a greater impact on environmental degradation and climate change than others, and to obscure this, along with the uneven effects it entails for different groups, is deeply problematic.[15] Failing to identify these crucial facets of the climate crisis moreover implies that the proposed solutions are likely to be inadequate, summoning forth a range of ethical considerations as well. Thus, far from denoting a stable and fixed geo-scientific signification, it seems more fruitful to

---

[12] P. J. Crutzen (2002) 'Geology of Mankind,' *Nature*, 415, 23.

[13] C. Eroukhmanoff and M. Harker (eds.) (2017) *Reflections on the Posthuman in International Relations: The Anthropocene, Security and Ecology*, E-International Relations.

[14] A. Burke et al. (2016) 'Planet Politics: A Manifesto from the End of IR,' *Millennium: Journal of International Studies*, 44(3), pp. 499–523.

[15] D. Faber, B. Levy, and C. Schlegel (2021) 'Not all People are Polluted Equally in Capitalist Society: An Eco-socialist Commentary on Liberal Environmental Justice Theory,' *Capitalism Nature Socialism*, 32(4), pp. 1–16.

consider how the 'Anthropocene' has been incorporated into contemporary security discourses.[16]

The 'Capitalocene' may well be 'an ugly word for an ugly system,'[17] but it has been useful in re-orienting discussions surrounding the human impact on the Earth System back to the global capitalist mode of production.[18] Jason Moore, a key proponent of the concept, affirms that the Capitalocene is not synonymous with capitalism per se; rather, the term denotes how capitalism operates as a 'World-ecology,' which, fuelled by an insatiable desire for appropriation, accumulation, and profit, puts nature to work in its own interests. According to Moore, global capitalism has ensured its own survival by exploiting 'The Four Cheaps' of food, labour, energy, and raw materials.[19] Yet, strictly speaking, it is *not* that capitalism is simply operating in a destructive manner *on* non-human nature. For Moore, and other advocates of the 'World-ecology' framework, such propositions would be erroneous for their purported reliance on, and reinforcement of, the untenable 'Cartesian dualism' of culture/nature which capitalism makes use of to exploit natural resources in the first place.[20] Alternatively, 'World-ecology' is derived from a Monist ontology, which posits the inextricable relations between culture and nature-hence, the use of composite terms such as 'capitalism-in-nature' or 'nature-in-capitalism.' In short, as a 'social' system, it would seem that capitalism cannot be considered independent from the biosphere of 'nature,' whether this be through abstraction or otherwise.

However, not all authors writing from a Marxist persuasion endorse the Capitalocene concept, nor do they profess satisfaction with the theoretical robustness of the World-ecology framework. On the contrary, John Bellamy Foster and proponents of eco-socialism (or 'ecological Marxism') have been quite scathing in their criticism of Moore. The main thrust of the charge is that efforts to avoid disassociating culture and nature through a Monist ontology constitute an intolerable departure from Marxist historical materialism and its foundational dialectical epistemology. The ontopolitical assumptions of Moore's work bring it into direct confrontation with prominent eco-socialists' own preference for a 'metabolic rift' theory to explain the relationship between capitalism and environmental degradation.[21] While

---

[16] M.J. Trombetta (2021) 'Security in the Anthropocene' in D. Chandler, F. Muller, D. Rothe (eds) International Relations in the Anthropocene, Palgrave Macmillan, pp. 155–172.

[17] J. Moore (2016) *Anthropocene or Capitalocene? Nature, History, and the Crisis of Capitalism*, Kairos, p. 5.

[18] As shall soon become apparent, articulating the human impact *on* the Earth systems would already, seemingly, fall foul to the pernicious culture-nature 'Cartesian dualism' which Capitalocene advocates are so keen to avoid.

[19] J.W. Moore (2017) 'The Capitalocene, Part 1: On the Nature and Origins of Our Ecological Crisis,' *The Journal of Peasant Studies*, 44 (3), pp. 594–630.

[20] Tracing the ideological roots of capitalism to the pillars of Western philosophical thought would suggest that the 'Capitalocene' is also a 'Westernocene'. A. San Román and Y. Molinero-Gerbeau (2023) 'Anthropocene, Capitalocene or Westernocene? On the Ideological Foundations of the Current Climate Crisis,' *Capitalism Nature Socialism*, 34(4), pp. 39–57.

[21] J. B. Foster (1999) 'Marx's Theory of Metabolic Rift: Classical Foundations for Environment Sociology,' *American Journal of Sociology*, 105(2), pp. 366–405.

Moore reproaches this theory for purportedly succumbing to 'Cartesian dualism' (i.e. through an unwarranted separation of 'culture' from 'nature'), Foster counters that such claims are rooted in an erroneous grasp of what Marxist dialectics actually is—i.e. that which both divides and unites in a totality at one and the same time.[22] This debate therefore takes on a decidedly ontological and epistemological character: How can we explain culture-nature relations, and are they separable, even at an abstract analytical level?

This brief foray into the onto-epistemological debate among Marxist-inspired authors serves to foreground two perceived limitations of the concept 'Capitalocene.' First, the seemingly stringent requirement to disavow any forms of 'Cartesian dualism' seems to preclude conceptions of the 'human' as an ontologically distinct and unique agential force.[23] This poses difficulties, in turn, when thinking the possibilities of human-led transformative change of the socio-ecological systems which are altering the Earth System. It is surely problematic if one cannot distinguish the ontological singularity of human beings as the agential capacities which they possess are clearly not commensurate with that of non-human beings or inanimate objects for that matter. Moore's account of transformative change flirts with a kind of economic determinism which, at times, seems hopelessly resigned to capitalism exhausting itself when 'cheap nature' becomes sufficiently depleted. Kees Jansen and Joost Jongerden ponder: 'Moore's singular monism, with capitalization and appropriation as the pivotal elements, seems to have but one direction: the approaching ending of the frontier (the 'epochal crisis' as the moment of change.'[24]

A related shortcoming of the World-ecology framework has been the under-theorization of the relationship between global capitalism and states. This anomaly is one which some proponents have, in fact, identified. Christian Parenti has stressed that states are indeed 'central' to discussions on the Capitalocene since they facilitate the delivery of non-human nature's 'use values' to capital, providing the necessary territorial, legal, and scientific/intellectual conditions within which capitalism must operate.[25] The theorization of capitalism-state relations is not only important for illuminating the role of the latter in facilitating the former, but also for discerning pathways for how human beings might transform them in ways which can restrict global capitalism and contain climate change. Moore himself has recognized the political import of this: 'the state is not only analytically central to the making of the capitalist World-ecology, but is the only institution large enough and powerful enough to allow for a progressive response to the escalating challenges of climate

---

[22] J. B. Foster (2016) 'Marxism in the Anthropocene: Dialectical Rifts on the Left,' *International Critical Thought*, 3, pp. 393–421.

[23] There are doubts over whether Moore's own work satisfies, or indeed can satisfy, this strict requirement. S. Nelson (2016) 'Book review essay—Sara Nelson on Jason Moore's "Capitalism in the Web of Life: Ecology and the Accumulation of Capital,' *Antipode Online*, 9th March, 2016.

[24] Jansen and Jongerden 'The Capitalocene response'.

[25] C. Parenti (2016) 'Environment-Making in the Capitalocene: Political Ecology of the State,' in Moore 'Anthropocene or Capitalocene?'; A. Battistoni (2023) 'State, Capital, Nature: State Theory for the Capitalocene,' Marxism and the Capitalist State, Springer.

change.'[26] It is to the relationship between global capitalism and states that we shall turn next.

International Relations and State-Capitalism Relations

Given the theoretical patrimony of the World-ecology framework, perhaps the most obvious place to begin exploring these questions is with World System Analysis (WSA).[27] This approach has had a significant impact on the field of International Relations (IR), particularly in the sub-field of International Political Economy (IPE), through its theorizing of the relationship between global capitalism and states. Diverging from classical accounts of IR, the central methodological novelty is that the world-system itself should be the primary unit of analysis to explain the roles and positions of individual states.[28] This 'world-system' is 'a much more integrated and interdetermined entity with a logic and structure of its own.'[29] The system itself, which essentially englobes the capitalist world-economy, determines the relative positions that states assume as its constituent parts. It is thus precisely economic processes such as a 'transnational division of labour' that shape the different roles that states fulfil. We thus arrive at a geographically differentiated and hierarchical typology that categorizes states as 'core,' 'periphery,' or 'semi-periphery.' The 'core' states, which include the most powerful capitalist states of Western Europe and North America, exist in an asymmetric and exploitative relationship vis-à-vis 'periphery' states. For example, states in the 'periphery' have historically assumed the specific role of providing raw materials and labour for the more advanced capitalist states in the 'core.' This amounts to a relationship of dependency since both the 'core' and the 'periphery' rely on each other; the result is an entrenched hierarchy which makes it difficult for 'periphery' states to rise out of their disadvantaged positions and to become part of the 'core.'[30]

The insights that WSA generates into capitalism-state relations would thus seem promising for a more comprehensive understanding of global capitalism's role in the climate crisis.[31] Yet enduring critiques continue to haunt its utility as a robust framework capable of illuminating with precision relations between global capitalism and states. One major issue is the problem of economic determinism. As Alexander Wendt carefully illustrated, this is a by-product of WSA's failure to break free from the

---

[26] Moore, 'Anthropocene or Capitalocene,' p. 10.

[27] A. Hornborg (2020) 'The World-system and the Earth System: Struggles with the Society/Nature Binary in World-System Analysis and Ecological Marxism,' *Journal of World-Systems Research*, 26(2), pp. 184–202.

[28] I. Wallerstein (1974) *The Modern World-System*, New York: Academic.

[29] C. K. Chase-Dunn (1979) 'Comparative Research on World-System Characteristics,' *International Studies Quarterly*, 23(4), pp. 601–623.

[30] Wallerstein, 'The Modern World-System'.

[31] Hornborg, 'The World-system and the Earth System'.

agent-structure problematic.[32] The reification of the capitalist world-system occurs as if it were an immutable backdrop within which states operate, meaning that it becomes difficult to explain the evolution of that system and the agency of states as its constituent parts.[33] As Wendt argues, it may be 'that the division of the world-system into three distinct structural positions (core, semi-periphery, periphery) is functional for the reproduction of capitalism, but this does not explain why the system developed that particular structure, nor does it guarantee that that structure will endure.'[34] This theoretical one-directionality of a system which determines the hierarchical positions of the state is compounded by another disposition which tends towards one-dimensionality. The calibration of states vis-à-vis the capitalist world-system takes place almost exclusively through capitalist economic processes, which, somewhat paradoxically for a historical sociological approach, limits the analytical potential for dealing with historical change. Prominent Marxist IR scholar Robert Cox was clearly not convinced: 'I am too much of a historian to accept a model like that other than as a tool, seeing where it works and doesn't work, not just accept it as a definition for world problems.'[35]

That there is an economic dimension to global capitalism-state relations is self-evident, but this is clearly not the only way that power relations are configured at the international or global scales. The kind of economic determinism which bedevils versions of WSA has rendered them vulnerable to constructivist critiques which place a methodological focus on ideas, identities, and discursive practices. Moreover, it follows that any comprehensive appreciation of state-capitalism relations would thus have to take stock of a more complex panorama, one which also includes racial, gendered and postcolonial dimensions.[36] Cox was also acutely aware of the problem of underdetermination through a disproportionate emphasis on economic relations, a premise which paves the way to an alternative approach to capture the complexity of relations at the international scale: 'we need to look at the material relations, ideas, institutions, production relations and world orders but not in any systematic form.'[37]

Without acknowledging *social* relations, it becomes not only difficult to account for how systemic change takes place, but also the roles that human beings can play in resisting and transforming world orders. The agential capacities of states, or other actors within states for that matter, become critically obscured. Cox, a self-confessed

---

[32] A. Wendt (1987) 'The Agent-Structure Problem in International Relations Theory,' *International Organization*, 41(3), pp. 335–370.

[33] S. Aronowitz (1981) 'A Metatheoretical Critique of Immanuel Wallerstein's "The Modern World System",' *Theory and Society*, 10(4), pp. 503–520.

[34] Ibid., pp. 347–348.

[35] R. W. Cox (2009) '"Overcoming the Blockage": An interview with Robert W. Cox,' Interviewed by Ana Saggioro Garcia and Miguel Borba de Sá, Toronto, p. 312.

[36] S. Feldman (2001) 'Intersecting and Contesting Positions: Postcolonialism, Feminism, and World-Systems Theory,' *Review* (Fernand Braudel Center), 24(3), pp. 343–371.

[37] R. W. Cox (2012) '"For Someone and the Some Purpose": An Interview with Robert W. Cox', in S. Brincat, L. Lima, and J Nunes (Eds) *Critical Theory in International Relations and Security Studies*, p. 7.

'pessimist,'[38] was nevertheless clearly not enamoured by the possibilities inherent in WSA for envisaging alternatives and mobilizing people to pursue transformative change. In contrast, a Gramscian approach, with its central concern being 'hegemony,' shows a greater promise for satisfying these crucial requirements. Since 'hegemony' and world orders are maintained through social relations, counterhegemonic movements can always emerge to challenge them.[39] Cox was quite clear that these human-driven forces must emerge from within states: 'in short, the task of changing world order begins with the long, laborious effort to build new historic blocs within national boundaries.'[40]

More recent Marxist-inspired research in IR has also placed a central focus on the social relations which bind together capitalism and states in non-deterministic terms. This work has turned around the concept 'global governmentality.' While Michel Foucault's original elaboration of 'governmentality' conceived it as a complex form of power oriented towards the government of populations—also known as the 'conduct of conduct'[41]—some authors have argued that at the international scale 'global governmentality' operates in a quite different way. Situated within a broader Marxist historical materialist framework, two primary claims have been advanced here. First, at the international scale, it is the hegemonic neoliberal variant of 'governmentality' which is most prominent and thus most of interest.[42] Western governments have promoted this principally, and it has then been disseminated by powerful liberal international institutions—e.g. the United Nations Development Program (UNDP) and the World Bank (WB)—despite the fact that the uneven nature of 'the international' makes its extension incredibly difficult if not impossible in reality.[43] Second, if there is a such a thing as neoliberal 'global governmentality,' then it is states and governments which are the immediate targets, and the ultimate aim is the restructuring and normalization of state-society relations in line with the requirements of global capitalism.[44] It is worth noting that these claims seem to undermine the very concept of 'global governmentality' as articulated by other 'Foucauldian' IR scholars since if the aim is to enact reforms by disciplining states and governments—e.g. through monitoring and rewards/punishments—then we would in fact be much closer to *disciplinary* power than anything else.[45] For authors like Jonathan Joseph, nevertheless,

---

38 R. Falk (2016) 'On the Legacy of Robert W. Cox,' *Globalizations*, 13(5), pp. 501–505.

39 R. Cox (1983) 'Gramsci. Hegemony and International Relations: An Essay in Method,' *Millennium*, 12(2), pp. 162–175.

40 Ibid., p. 174. See also R. Cox (1999) 'Civil Society at the Turn of the Millenium: Prospects for an Alternative World Order,' *Review of International Studies*, 25, pp. 3–28.

41 M. Foucault (1991) 'Governmentality,' in G. Burchell, C. Gordon, and P. Miller (Eds) *The Foucault Effect: Studies in Governmentality*, Chicago: University of Chicago Press.

42 J. Joseph (2012) The Social in the Global: Social Theory, Governmentality and Global Politics, Cambridge University Press, pp. 22–77.

43 Ibid.

44 Ibid.

45 J. Joseph (2009) 'Governmentality of What? Populations, States and International Organisations,' *Global Society*, 23(4), pp. 413–427.

this Marxist framing of 'global governmentality' allows us to explain not only the 'how' of governance but also, crucially, the 'why.'[46]

In addition, this Marxist framing of 'global governmentality' has important implications for how we consider capitalism-state relations within the context of progressive climatic change. Let us shift the focus now to international climate and environmental regimes of governance, which disseminate forms of 'green' or 'climate' governmentality such as 'resilience.'[47] While the use of 'resilience' discourses has clearly not been uniform semantically,[48] Joseph has argued that the variant being promoted through international institutions takes on a distinctly neoliberal complexion. For instance, corresponding policy frameworks advocate 'resilience' not only as a means to withstand external shocks, but also by promoting a reduced role for the state and the transference of responsibility on to individuals and communities through governmental technologies like 'self-reliance.'[49] Once again, the primary aim would appear to be the disciplining of states and national governments in line with global capitalism, which is deemed to be necessary before the effects percolate the agendas of state agencies, NGOs, local groups, and communities.[50] If we follow these claims to their natural conclusion, the diffusion of 'resilience' and other forms of green governmentality aim at bolstering or shaping state-capitalism relations in their neoliberal form, something which would run counter to any kind of transformative change in the face of the climate crisis.

Putting to one side momentarily the confluence of 'resilience' with neoliberal political rationality, some Foucauldian IR authors have raised objections to situating 'global governmentality' within a macro-theoretical Marxist framing.[51] While this kind of framework may well offer an answer to the 'why' of governance and not just the 'how,' the charge is that this comes at the cost of methodological incoherence as we stray too far from Foucault's own nominalist approach. For Hans-Martin Jaeger, Joseph takes for granted the 'deeper' ontological conditions within which governance activity occurs, but these are precisely 'the very phenomena that Foucault and governmentality studies have sought to denaturalize, namely, capitalism, the state and the states system.'[52] Scott Hamilton has expressed similar misgivings about the use of 'scientific realism' to ground governmentality inquiries: 'simply put, this is antithetical to the basic notion that governmentality is a nominalist mode of inquiry into

[46] Ibid.

[47] J. Stripple and H. Bulkeley (Eds,) (2014) *Governing the Climate: New Approaches to Rationality, Power and Politics*, New York: Cambridge University Press.

[48] P. Ferguson (2019) 'Discourses of Resilience in the Climate Security Debate,' *Global Environmental Politics*, 19(2), pp. 104–126.

[49] J. Joseph (2013) 'Resilience as Embedded Neoliberalism: A Governmentality Approach,' *Resilience*, 1(1), pp. 38–52; S. Davoudi (2016) 'Resilience and the Governmentality of Unknowns,' in M. Bevir (ed). *Governmentality after Neoliberalism*, New York: Routledge, pp. 210–249.

[50] Ibid.

[51] For a discussion see J. Busse and S. Hamilton (2021) 'Introduction' in J. Busse (Ed) *The Globality of Governmentality: Governing in an Entangled World*, London: Routledge.

[52] H. Jaeger (2014) 'Book Review: The Social in the Global: Social Theory, Governmentality and Global Politics,' *Acta Politica*, 49(2), p. 235.

fluid, unpredictable, and historically-contingent relations of power and thought.'[53] For all that, a crucial question remains: How might we retain coherence with the nominalist approach of Foucault's original groundbreaking analyses, whilst avoid slipping inadvertently into apologia for the neoliberal-capitalist project under scrutiny?[54]

The Global Capitalist Axiomatic and Its State 'Models of Realization'

As we have just seen, there is a varied Marxist IR literature to draw on for approximating the complex relationship between global capitalism and nation-states. In this section, we shall turn to the assemblage work of Gilles Deleuze and Felix Guattari, seeking to show how this framework can generate supplementary insights into this relationship. Of particular interest shall be in how states facilitate the operation of global capitalism and its disruptive tendencies towards the Earth System.

A Deleuzoguattarian ontology considers the world as being composed entirely of assemblages, which are complex multiplicities of both material and expressive elements.[55] As multiplicities, assemblages are not amenable to totalization, nor are they reducible to their constituent parts. While in the work of Deleuze and Guattari global capitalism and states comprise qualitatively different types of assemblage, given that they assume different forms and operate in divergent ways, they are also inseparable in that they exist in co-constitutive relationships. Before elucidating this non-deterministic relationship, let us first turn to how global capitalism operates as a unique assemblage.

Deleuze and Guattari argued that global capitalism operates as a specific type of assemblage known as an 'axiomatic,' which functions through incessant 'decoding.'[56] That is to say, it functions as a machine which uproots 'bodies,'[57] stripping them of their contextual significations, before recapturing and submitting them to a series of 'axioms.' This process is called 'axiomatisation,' whereby, simply put, 'axioms' are quantitative formulas or commands which treat their objects in a purely impersonal and functional way.[58] As Eugene Holland points out, global capitalism axiomatizes four kinds of flows: 'flows of liquid capital and abstract labour, of course,

---

[53] S. Hamilton (2014) 'Review: Add Foucault and Stir? The Perils and Promise of Governmentality and the Global,' *European Review of International Studies*, 1(2), p. 138.

[54] D. Chandler (2010) 'Globalising Foucault: Turning Critique into Apologia-A Response to Kiersey and Rosenow,' *Global Society*, 24(2), pp. 135–142.

[55] G. Deleuze and F. Guattari (2000) *Anti-Oedipus: Capitalism and Schizophrenia*, Minneapolis: University of Minnesota Press; G. Deleuze and F. Guattari (1987) *A Thousand Plateaus: Capitalism and Schizophrenia*, Minneapolis: University of Minnesota Press.

[56] Deleuze and Guattari, 'Anti-Oedipus.'

[57] I shall proceed with 'body' or 'bodies' in quotation marks to remind that we are in fact referring to assemblages composed of internal and external relations.

[58] Deleuze and Guattari, 'Anti-Oedipus,' pp. 34–35.

but also flows of raw materials and energy and of human populations.'[59] We have noted previously that in the World-ecology framework, capitalism's 'environment making' entails the appropriation and exploitation of 'Cheap Natures.' Translated to a Deleuzoguattarian lexicon we could say that these material 'bodies' are being decoded before being submitted to the axioms of the global capitalist axiomatic. Take, for instance, the concrete example of deforestation. Historically, indigenous populations of the Amazon have coded the forests in particular ways, such as in terms of a 'pachamama' worldview, which assigns them certain spiritual significations. Industrial-scale deforestation occurs as certain machines of global capitalism deterritorialize and decode the forests, uprooting and dissolving them of their contextual meanings, before submitting them to the axiom of 'general equivalence in the form of monetary value.' Logically, this implies that any struggle to preserve the forests might aim at re-asserting their indigenous meanings as 'life giver.'

Yet it is through another related process 'destratification' that we can grasp how global capitalism exercises a destructive impact on the natural world. As Kathryn Yusoff has argued, capitalism functions as a 'geosocial machine' that is 'feeding off the fossil stocks and mineral flows of the substratum.'[60] The exploitation of commodities like fossil fuels and the resultant discharge of massive quantities of greenhouse gases into the atmosphere disrupts the geological strata on which social-political assemblages themselves are dependent. Thus, the value of the concept 'geosocial strata' is in how it allows us to grasp the bi-directionality of the relationship between the 'social' and 'natural' worlds; that is, it is not just that capitalism is exploiting and disturbing a passive geophysical stratum, the destratification of the latter entails changes in the Earth System which then impacts upon the social systems themselves. It is thus deeply troubling that multinational energy corporations continue to rapaciously destratify the geophysical stratum at an alarming rate and with few signs of relenting.[61] One of the more ominous forms that this takes is 'carbon capital,' which can refer to how value is assigned by the market to the vast reserves of unexploited fossil fuels—the logical implication being that the use of such resources will occur, leading inevitably to further destratification and the exacerbation of climate change.[62] The global capitalist axiomatic's intrinsic and incessant orientation towards accumulation through the production of 'surplus-value of flow' drives its tendencies towards expansion and the maximization of profit, suggesting that there is theoretically no limit to the ravenous and destructive force that it can be.[63]

Thinking in terms of assemblages allows one to grasp how global capitalism does not operate on its own in its destructive impact on geological strata, but rather must

---

[59] E. W. Holland (2019) 'Market Theory and Capitalist Axiomatics,' *Deleuze and Guattari Studies*, 13(3), pp. 309–330.

[60] K. Yusoff (2017) 'Geosocial Strata,' *Theory, Culture, & Society*, 34(2–3), pp. 105–127.

[61] M. Taylor and J. Watts (2019) 'Revealed: The 20 Firms Behind a Third of All Carbon Emissions,' *The Guardian*, 9th October, 2019.

[62] J. Urry (2014) 'The Problem of Energy,' *Theory, Culture & Society*, 31(5), pp. 15–16.

[63] A. Toscano (2010) 'Axiomatic,' in A. Parr (ed), *The Deleuze Dictionary*, Edinburgh University Press: Edinburgh, p. 22.

also operationalize states and their mechanisms or 'models of realization.'[64] State assemblages are of a different type to global capitalism, though, comprising a divergent 'virtual' form. This means that 'actual' states are isomorphic with certain shared features—e.g. they are all centralized, hierarchical, and correspond with a particular territoriality (or 'milieu of interiority'). Moreover, unlike the immanent, impersonal, and functional way that the axioms of capitalism operate on decoded flows, state assemblages organize and regulate 'bodies' through 'coding.' Yet, it is how states facilitate the functioning of global capitalism, with the latter being dependent on the former, which is of most interest here. An assemblage approach theorizes capitalism as being in fact reliant on states, 'since it is the state which codifies the conditions under which the axiomatic can operate and regulate the flows of capital.'[65] State assemblages apply capitalism's axioms in particular spatial and material contexts, thus playing 'a crucial role in the regulation of capital movements and in the realization of surplus-value.'[66] Operating in these kinds of ways, states are therefore irrevocably implicated in global capitalism's exploitation and destratification of the geophysical strata, and, by extension, its altering of the Earth System.

Let us take the example of 'carbon markets' or 'carbon credits.' Once being hailed as a 'silver bullet' in the struggle against climate change, 'carbon markets' allow for the commodification of carbon, which, in turn, affords the most damaging polluters the chance to offset their emissions by purchasing 'carbon credits.' As Gareth Bryant puts it: 'climate change is produced through capital's drive to appropriate carbon by burning fossil fuels. Carbon markets commodify carbon by representing different forms of carbon appropriation as tradeable carbon allowances and credits.' Moreover, what interests us most here is that: 'states are central to both these relations, in delivering fossil fuels for capital and instituting carbon markets.'[67] Put in slightly different terms, the very carbon emissions which are the primary drivers of climate change are being subject to the axiom of 'general equivalence in the form of monetary value,' and states assume a key role in facilitating this exchange mechanism. Ultimately, the crux of the matter is that 'carbon markets' have been largely ineffective in stemming the tide of carbon emissions, confirming for many that the resort to market mechanisms to solve a problem fuelled by global capitalism is a futile enterprise.[68]

While all state assemblages are isomorphic in the ways that they provide the territoriality within which the axioms of global capitalism are applied, this does not, of course, mean that they are uniform. Variations in the concrete actualization of states proliferate, and as theorized in the world-system literature, their constitution is in an important sense relative to the positions that they assume vis-à-vis global capitalism. This means that they are certainly not commensurate in their roles and capacities

[64] Deleuze and Guattari, 'A Thousand Plateaus,' p4.58.

[65] P. Patton (2000) *Deleuze and the Political*, London: Routledge, p. 102.

[66] Ibid., p. 95.

[67] G. Bryant (2019) *Carbon Markets in a Climate-Changing Capitalism*, Cambridge: Cambridge University Press, p. 3.

[68] Ibid.

in facilitating global capitalism's rampant destratification of the geological strata; some states and societies logically harbour greater responsibility for climatic and environmental change than others, such as through their dependence on fossil fuels as exporters or importers.[69] The relative positions of states vis-à-vis global capitalism also translate into divergent socio-economic and security circumstances for their constituent populations, which, in turn, leads to widely divergent vulnerabilities in the face of progressive climate change.

State assemblages facilitate the global capitalist axiomatic, and thus its destratification of the geological strata, in another crucial way. They strive to regulate *human* flows within their territories, and they do so through heterogeneous configurations of power—for instance, through sovereign, disciplinary, and what is most pertinent here, governmental modalities.[70] As noted in the previous section, it is possible to view hegemonic forms of global governmentality as a means through which to discipline states and governments into adopting neoliberal reforms which align them with the requirements of global capitalism.[71] Thinking through assemblages permits a more nominalist inquiry into how a given state might adopt these hegemonic forms of governmentality (if at all), and/or how it might pursue alternative rationalities through state-coordinated regimes of governance.[72] Nonetheless, since all state assemblages facilitate the global capitalist axiomatic in some way, meaning that they are all 'capitalist,' a key concern becomes how they govern their populations in relation to climate and environmental change and in ways which serve the requirements of global capitalism. Of utmost concern here is how green governmentalities seek to constitute subjectivities. To approximate the complexities of such processes, let us turn to an author who has situated governmentality within a broader Deleuzoguattarian framework.

For Maurizio Lazzarato governmentality takes place at the intersection of two distinct but interrelated modalities of subjectivity: 'social subjection' and 'machinic enslavement.'[73] On the one hand, 'social subjection produces an "individuated subject"' through hegemonic systems of signification and subjectification which strive to regulate and code human 'bodies'; on the other, this mode of subjectification occurs in conjunction with another named 'machinic enslavement,' which is

---

[69] J. Depledge (2008) 'Striving for no: Saudi Arabia in the Climate Change Regime,' *Global Environmental Politics*, 8(4), pp. 9–35.

[70] B. Jessop (2007) 'From Micro-Powers to Governmentality: Foucault's Work on Statehood, State Formation, Statecraft and State Power,' *Political Geography*, 26, pp. 34–40.

[71] Reducing neoliberalism to capitalism would be a mistake. As Daniele Lorenzini puts it, neoliberalism constitutes 'a specific way or organising social relations on the basis of a rationality dictated by the market'. D. Lorenzini (2018) 'Governmentality, Subjectivity, and the Neoliberal form of Life,' *Journal of Cultural Research*, 22(2), pp. 154–166.

[72] R. Fletcher and J.A. Cortes-Vazquez (2020) 'Beyond the Green Panopticon: New Directions in Research Exploring Environmental Governmentality,' *EPE: Nature and Space*, 0(0), pp. 1–11.

[73] M. Lazzarato (2014) *Signs and Machines: Capitalism and the Production of Subjectivity*, Los Angeles: Semiotext(e), pp. 23–55. We can find these terms in Deleuze and Guattari's A Thousand Plateaus.

'superimposed on the production of the individuated subject.'[74] Machinic enslavement operates paradoxically through de-subjectification as 'bodies' are submitted to the non-representational axioms of global capitalism.[75] The expected result is that human beings are 'plugged in' to capitalism's productive machines as functional components or what Lazzarato (following Deleuze and Guattari) has called 'dividuals.'[76] The constitution of subjectivity through governmentality thus plays an important role in attempts to fix a quantitative relation of equivalence between the decoded flows of labour and the decoded capital used to purchase that labour.[77]

To illustrate how this theory of subjectivity can be of use in deciphering how states govern populations through green governmentalities let us return to hegemonic forms 'sustainable development' and 'resilience.' First, we might note how 'sustainable development' signals a novel way of restructuring the relations of capitalist production—e.g. through incorporating notions like 'green economy' and 'green growth.'[78] 'Resilience' has complemented this in environmental and development regimes of governance by seeking to constitute 'resilient subjects' in neoliberal terms as 'flexible,' 'adaptable,' 'self-reliant,' and 'entrepreneurial.'[79] In other words, the constitution of resilient individuals in these ways should facilitate their insertion as 'dividuals' into the machines of global capitalism; this should render them more amenable and equipped to deal with the vicissitudes of the market and the periodic shifts in relations of production in times of ecological crisis and environmental degradation. As Danny MacKinnon and Kate Derickson argue from a geographical perspective, 'resilient spaces are precisely what capitalism needs – spaces that are periodically reinvented to meet the changing demands of capital accumulation in an increasingly globalized economy.'[80]

Notwithstanding the affinity of 'resilience' with neoliberal governmentality, one cannot simply reduce the one to the other. For one thing, we might overlook a genealogy with roots in complex systems theory.[81] As 'resilience thinking' becomes incorporated into regimes of governance, the ontological underpinnings reveal a move away from Modernist frameworks separating the human from the non-human

74 Lazzarato, 'Signs and Machines,' p. 25.

75 Ibid.

76 Ibid., p. 27.

77 Deleuze and Guattari, 'A Thousand Plateaus,' pp. 452–453.

78 T. Wanner (2015) 'The New 'Passive Revolution' of the Green Economy and Growth Discourse: Maintaining the 'Sustainable Development' of Neoliberal Capitalism,' *New Political Economy*, 20(1), pp. 21–41.

79 For a discussion of whether resilience is an expression of neoliberal or post-neoliberal governance see L. Mavelli (2019) 'Resilience Beyond Neoliberalism? Mystique of Complexity, Financial Crises, and the Reproduction of Neoliberal Life,' *Resilience*, 7(3), pp. 226–228.

80 D. MacKinnon and K.D. Derickson (2012) 'From Resilience to Resourcefulness: A Critique of Resilience Policy and Activism,' *Progress in Human Geography*, 37(2), p. 254; see also R. Cretney (2014) 'Resilience for Whom? Emerging Critical Geographies of Socio-ecological Resilience,' *Geography Compass*, 8(9), p. 633.

81 J. Walker & M. Cooper (2011) 'Genealogies of Resilience: From Systems Ecology to the Political Economy of Crisis Adaptation,' Security Dialogue, 42(2), pp. 143–160.

world, and, in its place, we find assumptions on the complexity, contingency, emergence and non-linearity of inter-relations between them.[82] Thus, we might view resilience as an expression of 'postliberal' rationality, with important implications for both security and critique in a rapidly changing World. On the one hand, since unpredictability and vulnerability are deemed to be inherent features of these complex systems, resilience points towards self-reflexive adaptation as the most apt way of pursuing security.[83] On the other, resilience serves to delimit critique within certain parameters. As David Chandler has argued, resilience can 'transvalue' critique whereby it becomes a case of affirming complex relations (in pursuit of adequate adaptation) as opposed to transforming the system itself. [84] Resilient subjects are thus being constituted as passive and depoliticized; since the emphasis on adaptation does not address the systemic conditions which are constitutive of the climate and environmental crisis, this amounts to an acceptance of a permanent state of crisis and the renunciation of the right to security. Kevin Grove refers to this as a 'topological' form of security, which 'interiorizes the outside's de-territorializing potential' in order to maintain 'meta-stability'.[85] In the final analysis, whether we consider 'resilience' as the expression of global neoliberal or postliberal reason, the telos appears to be essentially the same: the maintenance of enduring yet differential state-capitalism relations with all the inequities and injustices that this entails.

Two final points emerge from an assemblage vantage point on state-global capitalism relations. First, while we have been adumbrating how state assemblages enable the global capitalist axiomatic, it does not follow that the former become entirely subservient to the latter. As David Bailey has pointed out, this could in fact have serious implications for states: 'if it goes too far towards the absolute deterritorialization that the capitalist axiomatic threatens to result in, the social system itself might become uncontrollable.'[86] This point evokes the paradox of how capitalist states are ultimately facilitating the destratification of the geophysical stratum, whereby the resultant climatic changes threaten the very capacities of these state entities to exercise sovereignty and provide security over their territories.[87] Yet this need not be the case. State assemblages do not coexist in a deterministic relationship with the global capitalist axiomatic, but a dynamic one; these entities can both add *and* subtract axioms, enabling *and* constraining global capitalism. Indeed, it is this dynamic relationship which puts the state front and centre in the struggle to mitigate

---

[82] D. Chandler (2014) Resilience: The Governance of Complexity, Oxford: Routledge.

[83] D. Chandler (2014) Resilience: The Governance of Complexity, Oxford: Routledge; B. Evans and J. Reid (2014) Resilient Life: The Art of Living Dangerously, Polity.

[84] D. Chandler (2014) Resilience: The Governance of Complexity, Oxford: Routledge.

[85] K. Grove (2017) Security beyond resilience, Environment and Planning D: Society and Space, 35(1): p184

[86] D.J. Bailey (2010) 'The European Rescue, Recommodification, and/or Reterritorialisation of the (Becoming-Capitalist) State? Marx, Deleuze, Guattari, and the European Union,' *Journal of International Relations and Development*, 13, pp. 338–339.

[87] This speaks to how resilience has also been subsumed within a national security discourse. See P. Ferguson, 'Discourses of Resilience in the Climate Security Debate,' Global Environmental Politics, 19(2), pp. 109–112.

global capitalism's damaging effects on the geological strata of the Earth. Second, the assemblage framework deployed here not only allows for a nuanced appreciation of state-global capitalism relations, but also, importantly, makes visible certain avenues for resistance and the transformation of capitalist states. Deleuze and Guattari were quite clear that neither global capitalism nor its state 'models of realization' can prevent the escape of decoded flows 'in all directions.'[88] Recalling Cox's appeal for a 'counterhegemonic' movement, how might we think this in terms of assemblages?

Capitalist States and Social Movement 'War-Machines'

From within IR there have been calls for a 'planet politics' to transcend the disciplinary limits of the discipline in search of new ways to tackle climate crisis.[89] The enduring inability of some international actors to take meaningful action, coupled with the global reach of the problem, suggests that such proposals are warranted. As some authors have cautioned, however, any 'planet politics' should not entail the re-formulation of the now discredited liberal cosmopolitanism or top-down global governance arrangements[90]; not only have these approaches failed to spur decisive climate action, but they've also been criticized for prioritizing Western interests. Any 'planet politics' must emerge from the bottom-up if it is to precipitate the kind of transformative change desperately needed.[91] If, as set-out in the previous section, states play a crucial role in both enabling and constraining global capitalism, then they logically become a key focal point over which the struggle to contain climate change takes place. Yet what this requires is more human agency and not less, and regrettably all too often the 'ontopolitics of the Anthropocene' seems to render intentional and transformative human agency somewhat obsolete.[92] A Deleuzoguattarian assemblage approach need not follow suit, it is just that any human agency is invariably situated and must take place from within assemblages.[93]

Deleuze and Guattari identified another type of assemblage which holds out significant promise for contesting and transforming capitalist states. The 'war-machine' is of a unique type or 'form,' and this has served it well in its historical opposition to states. Yet, just as there is no pure state assemblage—they are isomorphic, not

---

88 Deleuze and Guattari, 'A Thousand Plateaus,' p. 472.

89 A. Burke et al. (2016) 'Planet Politics: A Manifesto from the End of IR,' *Millennium*, 44(3), pp. 499–523.

90 D. Chandler, E. Cudworth, and S. Hobden (2017) 'Anthropocene, Capitalocene and Liberal Cosmopolitan IR: A Response to Burke et al.'s 'Planet Politics',' *Millennium*, 46(2), pp. 190–208.

91 Ibid.

92 D. Chandler (2018) *Ontopolitics in the Anthropocene: An Introduction to Mapping, Sensing and Hacking*, Routledge.

93 M. Robson (2024) 'Human Agency in the Production and Transformation of Global Security Assemblages,' *Millennium*, 52(2), pp. 384–407; see also S. Bowden (2020) 'Assembling Agency: Expression, Action, and Ethics in Deleuze and Guattari's *A Thousand Plateaus*,' *The Southern Journal of Philosophy*, 58(3), pp. 383–400.

identical—neither is there any 'pure' war-machine. It is also worth noting that the war-machine 'only entertains a potential or supplementary synthetic relation with war,' which results from its appropriation by the state: 'it is at the same time that the State apparatus *appropriates* the war-machine, subordinates it to its "political" *aims*, and gives it war as its direct *object*.'[94] Moreover, while it can seem like war-machines and state assemblages are self-contained and thus irreconcilable in their opposition to one another, in reality both coexist, being always intertwined in complex and dynamic relationships.[95] It is thus a question of tendencies and antipodal movements, as opposed to static and mutually exclusive configurations.

Variations in war-machine assemblages notwithstanding, they all comprise a shared abstract or 'virtual' form, which is advantageous for bottom-up social movements aiming to contest capitalist states and mitigate climate change.[96] While the state form is centralized and hierarchical, the war-machine is conversely a de-centralized and horizontal formation. It is an assemblage of the rhizome type, which is characterized by the dispersion of power and the proliferation of interconnections between its constituent 'bodies.' Unlike state assemblages who seek to maintain a 'milieu of interiority,' the war-machine incessantly strives to establish and maintain a 'milieu of exteriority.'[97] It is this radical exteriority and the relative autonomy that it awards, both in terms of space and forces, which can serve as the foundation for social movement war-machines aiming to contest and transform capitalist states.[98]

To maintain their power and autonomy, war-machine assemblages seek to elude the efforts by the state to capture and bring them within its territoriality. The aim is to cultivate a 'smooth space,' which is fluid, rhizomatic, and can serve as a political 'space of freedom' from which alternative political projects can be pursued. We have noted previously how the inculcation of resilience in regimes of governance can serve to restrict critique. The 'smooth space' of the war-machine, on the contrary, offers the promise of an external vantage point from which transformative critique and human agency can be mobilized. Social movements can thus forge these spaces as alternative material-affective-aesthetic 'atmospheres' or 'atmospherics' with potentially augmented capacities for human 'bodies' to act. [99] To achieve this, these formations draw or follow 'lines of flight' from state assemblages; this does not mean 'to flee,' as such, but rather 'to re-create or act against dominant systems of thought and social conditions.'[100] We might also think of this activity in Foucauldian terms

---

[94] Deleuze and Guattari, 'A Thousand Plateaus,' p. 420, original emphasis.

[95] Any distinction between the two types of assemblage must therefore be a 'formal' or 'virtual' one.

[96] The need for social movement war-machines is also recognized in E. W. Holland (2016) 'On/Beyond the Anthropocene,' *Deleuze Studies*, 10(4), pp. 564–573.

[97] Deleuze and Guattari, 'A Thousand Plateaus,' p. 354.

[98] A. Karatzogianni and A. Robinson (2009) *Power, Resistance and Conflict in the Contemporary World*, Routledge, p. 20.

[99] K. Grove (2017) Security beyond resilience, Environment and Planning D: Society and Space, 35(1), pp.184–194

[100] R. Deuchars (2011) 'Creating Lines of Flight and Activating Resistance,' *AntePodium: Online Journal of World Affairs*, n.p.

as 'counter conduct,' which signals a refusal to be governed in ways devised by the state.[101] One way that social movement war-machines can elicit 'lines of flight' and cultivate 'smooth space' is by targeting and 'dismantling' hegemonic forms of green governmentality like 'sustainable development' or 'resilience'—as we have noted, these essentially neoliberal (or postliberal) techniques seek to govern populations by constituting human subjectivities in passive ways which are geared towards maintaining a certain normalcy in state-capitalism relations.

Social movement war-machines can draw 'lines of flight' by exploiting the ambiguities or contradictions in forms of green governmentality. This begins by viewing hegemonic forms like 'sustainable development' or 'resilience' as: (i) irrevocably contingent or 'undecidable,' (ii) incapable of constituting human subjectivity decisively and definitively, and (iii) harbouring inherent transformative potential. Key signifiers like 'resilience' thus become a focus of political struggle, and by exploiting their immutable ambiguities social movement war-machines can evade state-led regimes of governance and subjectification, promote alternative and transformative notions, and ultimately produce a 'smooth space' from which to organize and project resistance against capitalist states implicated in climate change.

To illustrate how this might work in practice, we will continue with the example of 'resilience.' Through genealogical inquiry, authors have already shown how 'resilience' assumes a variety of different significations.[102] Embracing its contingency unearths the potential to conceive of human beings in less passive ways. If neoliberal/postliberal variants of 'resilience' advocating 'adaptability' and 'self-reliance' tend towards the production of passive and depoliticized 'resilient subjects'[103]—with the implied inevitability of liberalism-capitalism as the natural solution to our current plight—thinking it differently in terms of 'transformability' suggests that the transformation of deleterious socio-ecological relations generative of climate change would be possible.[104] We need not consider 'resilience' and 'resistance' as mutually exclusive concepts[105]; yet it is the corruption of the former in its hegemonic articulation that is in itself an act of resistance, since this operation can enable positive, liberating and transformative lines of flight.

As social movement war-machines seek to dismantle pernicious forms of green governmentality, the lines of flight that they pursue elicit elements which are vital for their functioning and make them powerful adversaries of capitalist state assemblages: pre-subjective desire, affects, and becomings. Let us now elaborate each of these, showing how they can increase the capacities of social movements to

---

[101] M. Foucault (2024) *What is Critique and the Culture of the Self*, The University of Chicago Press.

[102] See P. Bourbeau (2018) 'A Genealogy of Resilience,' *International Political Sociology*, 12(1), pp. 19–35. P. Bourbeau & C. Ryan (2017) 'Resilience, Resistance, Infrapolitics and Enmeshment,' *European Journal of International Relations*, 24(1), p. 6.

[103] Reid and Chandler, 'The Neoliberal Subject.'

[104] For an example see R. Cretney and S. Bond (2014) 'Bouncing Back' to Capitalism? Grassroots Autonomous Activism in Shaping Discourses of Resilience and Transformation Following Disaster,' *Resilience*, 2(1), pp. 18–31.

[105] In agreement with Bourbeau and Ryan, 'Resilience, Resistance, Infrapolitics.'

contest and ultimately transform the capitalist states implicated in climate change. We will refer once more to 'resilience' for explanatory purposes.

In the work of Deleuze and Guattari, 'desire' is the primary constitutive force of any socio-political assemblage. That is, it is desire which produces, circulates, and takes root in these arrangements. Nonetheless, state assemblages seek to regulate, restrict, and channel human desire in particular ways. For instance, as we have touched on earlier, this takes place through different modalities of power, and in relation to climate change forms of green governmentality such as 'sustainable development' and 'resilience.' Thus, it is not merely that these governmentalities aim to govern populations by constituting human beings in particular ways—e.g. as 'resilient subjects'—but that they also restrict and channel the flow of human desire towards the reproduction of state-capitalism and socio-ecological relations. By dismantling signifiers like 'resilience' and promoting alternative transformative significations, social movement war-machines can free lines of flight and effectuate a release of pre-subjective desire in the process. The release of this inherently productive force, in turn, effectuates a proliferation of rhizomatic connections among human beings, along with concomitant increases in their capacities to act. The more connections which are established between 'bodies,' the greater the intensity and capacities of these 'bodies' to act in their collective efforts to contest capitalist state assemblages and contain climate change. We can view these operations, therefore, as acts of liberation and resistance.

By promoting transformative notions of signifiers like 'resilience' social movement war-machines can also bring about a release and diffusion of 'affects.' Broadly speaking, 'affect' denotes a relational force between 'bodies,' which corresponds with an increase/decrease in their capacities to act.[106] Let us take, for instance, the example of 'hope.' We have already noted that hegemonic notions of resilience, by asking people merely to adapt to the world, engender passivity and depoliticization in human beings. In this way resilience seeks to 'interiorize affective potential'.[107] 'Apocalyptic' narratives which sometimes accompany can impel feelings of despair and hopelessness, diminishing the capacity of 'bodies' to act.[108] Alternatively, invoking transformative connotations of 'resilience' can bring about more powerful affective relations, restore feelings of 'hope,' which social movement war-machines can then circulate as affects.[109] As an 'affect' hope can increase the capacity of human beings to act within these arrangements, and thus may even constitute the most significant force in the struggle to contain climate change.[110]

---

[106] Deleuze and Guattari, 'A Thousand Plateaus'.

[107] K. Grove (2018) Resilience, Oxon: Routledge, 272.

[108] M. Fagan (2017) 'Who's Afraid of the Ecological Apocalypse? Climate Change and the Production of the Ethical Subject,' *The British Journal of Politics and International Relations*, pp. 1–20.

[109] See also D. Chandler (2023) 'Rethinking 'Hope' and 'Resilience' in the Anthropocene: An Interview with David Chandler,' *World Futures Review*, 15(1),

[110] For an understanding of hope as affect see B. Anderson (2006) 'Becoming and Being Hopeful: Towards a Theory of Affect,' *Environment and Planning D: Society and Space*, 24(5).

Finally, dismantling hegemonic forms of green governmentality effectuates a series of 'becomings.' A complex notion in the work of Deleuze and Guattari, 'becoming' denotes 'continuous transformation,' and, crucially, it is through which that social movement war-machines can elide the pernicious gendered, racial and political hierarchies historically erected by state assemblages. While neoliberal/postliberal 'resilience' works to curtail the 'becoming-otherwise' of life, [111] 'dismantling' this signifier can effectuate a series of becomings which can increase the capacities of social movement war-machines. There are at least three 'becomings' which can serve these formations in their struggle against capitalist state assemblages.[112] First, while the neoliberal/postliberal variant of 'resilience' typically expresses a gendered phallocentrism through its advocacy of notions like 'self-reliance,'[113] invoking more transformative notions can beget a 'becoming-woman' that brings into play new affective interrelations among 'bodies' characterized by empathy, compassion, and care.[114] Second, whereas similarly hegemonic notions of 'resilience' remain tied to an anthropocentrism whereby the 'resilient subject' is situated as master of its own fate—even when paradoxically restricted by the complex systems within which it is immersed and must adapt to—[115] the effectuation of a 'becoming-animal' embraces multiplicity, heterogeneity, and the radical interconnectedness of life with all of its power and transformative potential. Third, contra the depoliticization induced through neoliberal/postliberal notions of 'resilience' social movement war-machines can instead engender a 'becoming democratic,' which could pave the way to greater public participation in decision-making and potentially alternative socio-ecological arrangements.[116]

If social movement war-machines are to be successful in transforming capitalist states, contestation at the level of discourse alone will most likely not suffice. Some form of performative political action will also be necessary.[117] Being horizontal,

---

[111] K. Grove (2018) Resilience, Oxon: Routledge

[112] Deleuze and Guattari speak of many more 'becomings,' although a more sustained engagement is beyond the scope of this chapter.

[113] M. Sawer (1996) 'Gender, Metaphor and the State,' *Feminist Review*, 52(1), pp. 118–134; see also J. McAuliffe, 'She's Making Profit Now: Neoliberalism, Ethics, and Feminist Critique,' *philoSOPHIA*, 10(1), pp. 24–46.

[114] E. Allison (2017) 'Toward a Feminist Care Ethic for Climate Change,' *Journal of Feminist Studies in Religion*, 33(2), p. 154.

[115] S. Wakefield, D. Chandler & K. Grove (2021) 'The asymmetrical anthropocene: resilience and the limits of posthumanism,' *Cultural Geographies*, 29(3), p. 11; see also N. Clark (2011) *Inhuman Nature: Sociable Life on a Dynamic Planet*, London: Sage.

[116] While some have questioned Deleuze's democratic credentials, others have offered an alternative reading. See, for instance, P. Patton (2008) 'Becoming-Democratic,' in I. Buchanan and N. Thoburn (Eds) Deleuze and Politics, Edinburgh University Press.

[117] E. Swyngedouw and H. Ernstson (2018) 'Interrupting the Anthropo-obScene: Immuno-biopolitics and Depoliticizing Ontologies in the Anthropocene,' *Theory, Culture & Society*, 35(6), pp. 3–30; see also.

W. E. Connolly (2015) 'Confronting the Anthropocene and Contesting Neoliberalism: An Interview with William E. Connolly: Interviewer: Bradley J. Macdonald,' *New Political Science*, 37(2), p. 259–275.

dispersed, and highly mobile formations, war-machine assemblages are particularly adept at carrying out this kind of action which can degrade the capacities of capitalist states. Let us take the example of the 'general strike,' which William Connolly has advocated as a means through which to engage in political struggle in the 'Anthropocene.'[118] When carried out by social movement war-machines a general strike can, on the one hand, increase the capacities of its constituent 'bodies' to act through combined action and by inspiring people through this expression of dissent. This kind of performative action can also decrease the capacities of states and capitalist machines due to the detrimental effects that it can have on their functioning, productivity and profit. By degrading the capacities of capitalist states in these ways overwhelming pressure can be put upon governments to act or possibly even reform the very structure of the state to augment democratic participation.[119]

Moreover, social movement war-machines formations can also make use of various kinds of 'weapons.' A further caveat is in order here since the 'weapons' of which Deleuze and Guattari speak do not connote any kind of violence, but, rather, are understood as 'affects' which bring about increases/decreases in the capacities of 'bodies' to act.[120] Information and communications technologies (ICTs) can be understood in such terms and are at the disposal of social movement war-machines. ICTs can increase their organization, coordination, and communication capacities, which, when combined with direct action such as protests or strikes, can be a potent force in degrading the capacities of capitalist state assemblages. They can also transmit the 'framing' of the problem of climate change, which again is essential for disseminating affects such as hope. This leads to the expansion, proliferation, and intensification of rhizomatic connections between constituent 'bodies' of these social movement war-machines. Of course, climate activists are well aware of the utility of social media platforms in building and directing communities through engaging and affective narratives.[121]

Conclusions

This chapter has aimed to address some perceived limitations in how the concept 'Capitalocene' has been used in the academic literature by turning to the assemblage work of Deleuze and Guattari. A Deleuzoguattarian assemblage framework offers a unique vantage point from which to study global capitalism's constitutive role in climate and environmental change, since it permits a welcome analytical purchase

---

[118] Ibid., p. 267.

[119] Extinction Rebellion's demand for 'citizens' assemblies' is one possible objective of this kind of direct action. Extinction Rebellion (2019) 'The Extinction Rebellion Guide to Citizens' Assemblies,' *XR-UK Citizens' Assemblies Working Group*, 25th June 2019.

[120] Deleuze and Guattari, 'A Thousand Plateaus,' p. 400.

[121] E. Wielk and A. Standlee (2021) 'Fighting for Their Future: An Exploratory Study of Online Community Building in the Youth Climate Change Movement,' *Qualitative Sociology Review*, 17(2), p. 23.

on the crucial relationship between global capitalism and states. It is first worth re-iterating that this is a dynamic non-deterministic relationship as states have the capacity to both enable and constrain global capitalism; it follows that states would thus become a key focal point in the struggle to contain climate change. Not only can we generate insights into how states facilitate global capitalism's destratification of the geophysical stratum on which it feeds, but also, importantly, this permits a nominalist inquiry into how states govern populations in ways which re-produce a certain normalcy in global capitalism-state relations. States can all be considered capitalist in so far as they apply the axioms of global capitalism through territoriality, but as assemblages or dynamic multiplicities they do not govern their populations in uniform ways vis-à-vis climate and environmental change. Mapping the heterogeneous forms of green governmentality which states mobilize to govern their populations is thus a necessary first step in opening critical avenues for socio-political resistance and the transformation of capitalist states into more benign configurations. This chapter has also provided a brief sketch of what social movement war-machines might look like, and how they might pursue these aims.

I would like to make some final comments on these findings and their implications. While this chapter has underlined the potential of social movement war-machines for the contestation and transformation of capitalist states, these formations face formidable obstacles in practice. Establishing and maintaining connections between the constituent nodes requires considerable effort, and their threat to capitalist states ensures that reactionary forces will inevitably mobilize against them. This will logically entail attempts to subsume the antagonistic formations back within the territoriality of states. It is also important to sound a note of caution. Historically speaking, it is clear that war-machine formations have not always assumed a peaceful orientation. Social movements run the risk of incurring significant costs, not least that of alienating the general public, if they abandon peaceful means of bringing about the positive and stable transformation of capitalist states towards the ultimate aim of containing climate change. It is also worth emphasizing that the 'lines of flight' pursued by war-machines do not always turn out well. Being essentially lines of deterritorialization, prudence requires that one proceed with caution, striving for incremental change by re-connecting and re-combining 'bodies' in productive ways.[122] As Nicolas Tampio has argued, Deleuze's vision of 'revolutionary' assemblages was in fact much more modest than some assume, whereby 'the means of political transformation should be exercised cautiously and experimentally, and the end of politics is a more free and egalitarian political order.'[123]

---

[122] Patton, 'Deleuze & the Political,' p. 107.

[123] N. Tampio (2009) 'Assemblages and the Multitude: Deleuze, Hardt, Negri, and the Postmodern Left,' *European Journal of Political Theory*, 8, pp. 383–400.

Chapter 8
Conclusions

Matthew Robson

Since its emergence in the natural sciences, the 'Anthropocene' has triggered a quite remarkable volume of work in academic fields like International Relations (IR), particularly since its central claim is something we have been cognizant of for some time: human beings are responsible for altering the planet through their disruptive activities. Debates have taken place on the implications and suitability of the concept to denote our contemporary socio-ecological context, alongside other important work examining the 'ontopolitics of the Anthropocene.'[1] As stated in the Introduction section, an analytical focus on 'ontopolitics' reflects a desire to interrogate the 'often-unrecognized ontological assumptions which underpin political theories, practices, modes of thought and governmental mechanisms.' Some authors have raised concerns regarding the ontopolitical assumptions of some approaches to the 'Anthropocene,' which seemingly resort to an 'affirmation' of complexity and contingency, along with the dissolution of the human subject and concomitant attenuation in human agential capacities.[2] These are very serious charges indeed since any hope of addressing the ensuing ecological crisis requires more human agency and not less.

This edited volume has sought to explore the ontopolitics of the 'Anthropocene' by bringing together some of its different framings and conceptualizations, showing how authors make use of them in IR. Together these original contributions lend

[1] D. Chandler (2018) *Ontopolitics in the Anthropocene: An Introduction to Mapping, Sensing and Hacking,* Routledge.

[2] Ibid.

M. Robson (✉)
Faculty of Political Science and Public Administration, Chiang Mai University, Chiang Mai, Thailand
e-mail: matthew.r@cmu.ac.th

M. Robson (ed.), *The Anthropocene, Ontopolitics, and International Relations*,
https://doi.org/10.1007/978-981-96-2925-1_8

weight to existing research illustrating that how we frame our contemporary socio-ecological context most certainly matters.[3] All the authors have expressed an interest in how the social/cultural and the natural worlds, or the human and the non-human, relate to one another. Most have done so through framings of the 'Anthropocene' derived from relational ontologies, though not all. This reflects the now prevalent view that the 'Anthropocene' critically undermines the binary oppositions that have structured the Modernist pursuit of knowledge in IR. In this way, we might say that the authors are responding to the ontological questions posed by the 'Anthropocene,' which leads them to a mapping and rethinking of the relations between society/nature and the human-non-human as a way of conducting political critique and formulating novel political proposals.

In this final Conclusions chapter, I shall carry out a sustained analysis of the ontopolitics of the different framings of the 'Anthropocene' found in this volume. I will organize my reflections around how the authors' ontological assumptions inform the political possibilities for knowledge production/critique, human subjectivity/agency, praxis, and transformation. As we shall see toward the end of the chapter, these ontological commitments also have a bearing on the kind of ethics that the authors pursue. I aim to show that the chapters reveal a varied ontopolitical panorama, with divergences in two main directions: first, in terms of the authors' political claims and proposals, and second, in terms of the implications for political subjectivity and transformative human agency. Thus, it is not that the 'Anthropocene' (variously defined), or its 'posthuman' relational framings per se, lead inevitably to the dissolution of the human subject and depoliticization; rather, it is our own ontological assumptions vis-à-vis relationality or Monism which would be the determining factor. The principal argument developed in this chapter is that unremitting engagement with the ontopolitics of the 'Anthropocene' (broadly conceived) is essential to fulfill the promise of critical inquiry and forge alternative political imaginaries in these times of ecological crisis.

Ontopolitics of the 'Anthropocene'

In their chapter, Barros-Platiau et al. invoke a more positivist conception of the 'Anthropocene,' which is grounded on a Modernist ontology. As such, the chapter displays a clearer ontological distinction between the social/cultural and natural worlds than the others, which, from some viewpoints, could be considered anthropocentric. The natural world seems to retain its status as the backdrop or stage on

[3] Ibid.; E. Randazzo and H. Richter (2021) 'The Politics of the Anthropocene: Temporality, Ecology, and Indigeneity,' *International Political Sociology*, 15, pp. 293–312; F. Chipato & D. Chandler (2023) 'The Black Horizon: Alterity and Ontology in the Anthropocene, *Global Society*, 37(2), pp. 157–175; E. Randazzo and H. Richter (2024) 'Political Action in Planetary Times: Extinction Activism, Anthropocene Ontopolitics, Indigenous Complexities,' *Political Geography*, 112, 103107, pp. 1–10.

which human affairs are being played out. From these onto-epistemological foundations, the authors place a strong emphasis on human beings both as the constitutive force driving climate and environmental change, and as those who must devise innovative solutions to the ensuing crisis. Thus, ontopolitically, the Modernist framing of the 'Anthropocene' paves the way to political proposals which are essentially technocratic in their constitution, being ultimately a reformed and improved top-down global governance and diplomacy architecture fit for the new security context.

Alternatively, Pereira et al. subscribe to a 'more-than-human' framing of the 'Anthropocene.' Mapping culture/nature or human/non-human relations thus becomes an ontopolitical priority for the authors, and this prompts them to embark upon an exploration of the Indigenous onto-epistemologies and practices of the Amazon. It is the shared ontological commitment to relationality which bridges post-Anthropocentric IR and the Amazon, and which suggests that the turn to Indigeneity can open up novel political possibilities. For the authors these revolve around 'land governance and the bioeconomy,' the 'transnational politics of freshwater ecosystems and hydrological connectivity,' and even those related to the 'risks and threats of environmental militarization.' Indigenous knowledge and practice are thus seen as harboring important resources to advance post-anthropocentric IR, but, as the authors insist, this ontopolitical dialogue must be bi-directional as Western ontologies and epistemologies can also be a source of insights for political practice in the Amazon.

Kuetting is similarly interested in rethinking and reformulating relations between the social/cultural-natural worlds, but it is 'multi-species relations' which are of primary concern. For the author, the crux of the problem with these relations at present is their commodification, understood to be rooted in the epistemological separation of culture from nature that can be traced back to the European Enlightenment and colonization. Kuetting's ontopolitical assumptions about the 'Anthropocene' thus seem to spur the critique of the 'binomic ordering' of current multi-species relations, and the desire to reconceptualize them in more environmentally sustainable and ethical ways. With regard to the latter the chapter surveys a range of scholarship, from Indigenous/decolonial frameworks (which appear to share similar ontopolitical assumptions with the author), 'environmental justice' literature (some of which deemed to be beset by a restrictive anthropocentrism), and that which concerns animal rights. Ultimately, the author concludes that although 'rights-based approaches emulate the liberal, individualistic roots of a binomic system, they are nevertheless the most promising avenue for multi-species representation in the Anthropocene.'

Molinero-Gerbeau and Perez's explanation of our contemporary ecological crisis shares some similarities with that of Kuetting's, but, through a preference for the term 'Capitalocene,' there is more of an explicit emphasis on global capitalism as the principal driver. The World-ecology framework advocated is grounded on a relational ontology whereby the social/cultural and natural worlds are once again deemed to be irrevocably interconnected, but within this 'Web of life' it is global capitalism which is the all-encompassing constitutive force. Indeed, global capitalism makes use of 'dualisms' like society/nature to appropriate and exploit nature or 'cheap nature.' Within this particular relational framing, which permits a contentious kind of

Monist dialectics,[4] the ontopolitical task becomes primarily one of critique, tracing, and dismantling these pernicious 'dualisms.' Taking international migration as the analytical focus, the authors aim to deconstruct categories like 'climate migrant' on the basis that they are expressions of two dualisms erected by the capitalist world ecology to regulate human mobilities: (society/nature and legitimate/illegitimate). The ontopolitics of the 'Capitalocene' foregrounds not only this deconstructive moment but also a reconstructive one as the authors propose the alternative term 'CoPe' (Core-Periphery) as a more 'holistic' way of capturing the phenomenon of mobility in the Capitalocene.

In a similar vein, San Román's chapter is concerned with mapping how culture-nature relations are configured in our present socio-ecological context, but, in this case, the political contribution remains at the level of critique. For the author, conceptions of the 'Anthropocene' *and* the 'Capitalocene' turn out to be unsatisfactory since they both rely on different untenable anthropocentric suppositions. These relate to uncritical conceptions of 'technology,' which fail to account for the 'dualistic' premises on which it is typically conceived. This also suggests that both the 'meta-narratives' of the 'Anthropocene' and 'Capitalocene' ultimately fail to live up to the Monist ontology which many of their adherents espouse. For San Román, a proper mapping of social/cultural-natural relations reveals that 'technology is the problem.' It is technology that erects the 'dualism' separating 'humanity from Nature and vice versa,' and taking as its principal orientation the domination of Nature. Nonetheless, importantly, 'technology' is not being conceived in purely materialist terms, but as a 'worldview' or 'hegemonic cultural system' that derives from Western knowledge. Thus, once again, the ontopolitical assumptions of the author lead to a mapping of culture-nature relations, but the seemingly more stringent commitment to Monism foregrounds a rejection of both the 'Anthropocene' and the 'Capitalocene,' and the advocacy of the 'Techno-Westernocene' as the most appropriate designation of our contemporary socio-ecological context.

In the final stand-alone chapter, Robson takes the 'Capitalocene' as a point of departure but ultimately eschews any concrete term for our contemporary socio-ecological context.

Like previous chapters, however, his ontopolitical assumptions lead to what is deemed to be a necessary mapping of contemporary culture-nature relations. Deploying a Deleuzoguattarian 'assemblage' framework Robson aims to show how states facilitate global capitalism's destratification of the geological strata. Seemingly, an important part of this is the government of populations through various neoliberal/postliberal forms of 'green' or 'climate' governmentality, which typically aim to re-produce state-capitalism relations in their current form with all the climate and environmental implications that this entails. Shining light on these relations, it is argued, is necessary to not only engage in the political critique of how the state-capitalism nexus exploits and degrades the natural world, but also for an appreciation of how the constitution of human subjectivity takes place. This is articulated

[4] A. Hornborg (2020) 'Dialectical Confusion: On Jason Moore's Posthumanist Marxism,' *Historical Materialism*, 25th June, 2020.

as a necessary first step for the political contestation of capitalist states through the formation of social movement 'war-machines'.

As we have seen, most of the authors in this volume have embraced relational ontologies of one stripe or another for framing our contemporary socio-ecological context. The ontopolitical orientation of the chapters toward critically mapping the relations between the social/cultural and natural worlds seems to confirm a tendency to privilege the 'is' of the World over the normative 'should be' typical of Modernist frameworks.[5] Yet, I do not wish to characterize these contributions as a celebration of the 'Anthropocene.' The authors invariably express their concern and dissatisfaction with present socio-ecological relations, and with certain aspects of how our contemporary international political and economic structures are implicated. This, in turn, paves the way for the articulation of a range of different political proposals. Thus, we could argue that the authors express a certain desire to go beyond the 'is' of the World through a contemplation of the 'could be,' 'should be,' or 'must be'—whether this be through the construction of alternative governance regimes, the rights for non-human sentient beings, or the pursuit of greater democratic participation. It is just that the ontopolitical commitments inherent in these relational framings of the 'Anthropocene' imply the necessity of first critically mapping the complex inter-relations between society/culture-nature and the human/non-human before any political proposals can be formulated. An exploration of the 'is' of the Word becomes necessarily connected with the articulation of the normative 'ought to.' It thus seems more apt to characterize these contributions as a collection of attempts which aim to deal with the onto-epistemological implications of the 'Anthropocene' for the field of IR. This becomes a question of methodological coherence more than anything else, whereby the authors commit to working within certain, albeit divergently defined, relational parameters.

In addition, that the political proposals being made by some authors emerge from the mapping and critique of existing culture-nature relations need not *necessarily* be considered a shortcoming. It is just accepted (implicitly in most cases) that this activity must emerge immanently from the relational structures within which human beings are immersed. Through the authors' advocacy of relational approaches it is difficult to imagine how this could be any other way, and, arguably, as the contributions of this volume seem to have shown, this need not shut down the possibilities for political critique, contestation, or the formulation of alternative political proposals. In Michel Foucault's work, for instance, the human subject finds itself inextricably bound up with relations of power, meaning that the possibilities for resistance or 'counter-conduct' can only be envisaged immanently from within these relations.[6] Similarly, in the work of Gilles Deleuze and Felix Guattari, one finds an unmistakable affinity with lines of deterritorialization (albeit with qualification), and yet

---

[5] Chandler, 'Ontopolitics in the Anthropocene,' p. 16.

[6] M. Foucault (1982) 'The Subject and Power,' *Critical Inquiry*, 8(4), pp. 777–795; M. Foucault (2007) Security, Territory, Population: Lectures at the College de France, 1977–1978, Palgrave Macmillan; M. Foucault (2023) 'What is Critique?' In H.P. Fruchaud and D. Lorenzini (Eds.) *What is Critique? & The Culture of the Self*, Trans. C. O'Farrell, The University of Chicago Press.

any human activity oriented toward this must invariably take place immanently from within socio-political assemblages.[7] Some chapters in this volume display a similar immanent deconstructive ethos, counter-conduct, or 'critical attitude,' even if they do not articulate it in these terms; whether it be in the de/reconstruction of contemporary categories of migration governance (Molinero-Gerbeau and Pérez), the critical interrogation of existing liberal rights regimes (Kuetting), or the contestation of neoliberal/postliberal oriented 'green' or 'climate' governmentality (Robson).

For all that, to address adequately any charges of merely 'affirming' the 'Anthropocene,' which in David Chandler's terms would be an affirmation of the complexity and contingency of 'what exists rather than seeking to go beyond this,'[8] the onus must shift towards how alternative worlds can be pursued. It becomes necessary to gauge to what extent the ontological assumptions grounding these different framings of the 'Anthropocene' allow for the political contestation of subjectivity and the agential capacities of human beings to take transformative action. It is perhaps worth underlining that we have not conceptualized the volume specifically as an inquiry into the transformative possibilities inherent in different framings of the 'Anthropocene.' It is not even immediately apparent that all the authors would advocate such a thing in the first place—some of the political proposals advanced could perhaps be accommodated within existing political and economic structures, although in other cases this seems unlikely. Nor can we assume that 'transformation' means the same thing to each of the contributing authors. Nonetheless, all chapters at least *imply* the need for some kind of substantive change in international political and economic structures, whether this be through changes to global governance, international migration regimes, or the operation of global capitalism. Thus, I shall assess how the ontological assumptions in different framings of the 'Anthropocene' harbor divergent implications vis-à-vis the potential for dissident political subjectivities and transformative human agency. For that we will bring the chapters into dialogue (and supplement them) with other work in the field in an attempt to move discussions forward and tentatively suggest lines of future inquiry.

Let us begin once more with the chapter by Barros-Platiau et al. As noted, the authors employ a more Modernist framing of the 'Anthropocene,' and it is through which they make the case for novel governance and diplomacy structures to deal with contemporary climate-cyber threats. Ontopolitically speaking, the separation of the human subject from the natural world does not necessarily preclude the kind of transformative agency which many deem necessary to address the climate crisis. Nonetheless, there are other assumptions which would appear to cast doubt on this. First, the conception of the 'Anthropocene' articulates an undifferentiated 'anthropos,' which does not specify who or what is ultimately responsible for the climate crisis, nor which groups are more vulnerable as a result. If we do not assign responsibility, the kind of transformative human agency necessary to contest power may also remain opaque. Second, the ontopolitical commitments of this Modernist framing allow for

[7] G. Deleuze and F. Guattari (1987) *A Thousand Plateaus: Capitalism and Schizophrenia*, Minneapolis: University of Minnesota Press, pp. 186–191.

[8] D. Chandler, 'Ontopolitics in the Anthropocene,' p. 183.

the constitution of the human being as both subject and object of security practices. Yet as some authors within IR have argued, the implications of the 'Anthropocene' expose these kinds of security discourses for failing to account for the relations between the human and non-human. According to Audra Mitchell, for instance, it is the 'whole world' that is now being threatened through alterations in the Earth System, which we can understand as 'irreducible, heterogeneous forms of collective being.'[9] Any proposals to secure human beings through top-down technocratic measures also seem to only raise further questions about the effectiveness of liberal global governance in tackling the drivers of climate change.[10]

In the chapter by Pereira et al. the 'more-than-human' relational ontology grounding their conception of the 'Anthropocene' does not lead to the complete dissolution of the human subject or human agency. This is most evident in the authors' exploration of the Indigenous knowledge and practices surrounding 'transformative governance.' As the authors note, 'in the north-western region of the Amazon, Indigenous groups often govern their shared territories collectively, guided by ecological principles that subordinate political and social practices to the cycles of nature.' Moreover, it is not necessarily the case that human subjectivity and agency become solely reactive and adaptive to the World; transformative change is articulated as a possibility through recognition that Indigenous experiences 'could yield valuable insights for reconceptualizing world politics beyond modernist, conventional economic-growth-driven capitalist and techno-managerial approaches to human/non-human relations.' The shift of focus to Indigenous frameworks in search of resources to reconceptualize situated forms of human subjectivity and agency within relational ontologies is something we have seen elsewhere in the field of IR.[11] To make these claims it is necessary to avoid the complete dissolution of culture/nature or human/non-human binary distinctions, and embrace what we might refer to as a kind of 'weak Monism' that still leaves room for intentional, even transformative, human agency. Further research in this area could be fruitful in fleshing out further what this might look like.

As mentioned Kuetting's chapter explores various avenues for re-conceiving multi-species relations in pursuit of multi-species justice, but the role of human beings in realizing this goal remains somewhat unclear. If the crux of the problem

---

[9] A. Mitchell (2014) 'Only human? A worldly approach to security,' *Security Dialogue*, 45(1), pp. 5–21.

[10] D. Chandler, E. Cudworth, and S. Hobden (2017) 'Anthropocene, Capitalocene and Liberal Cosmopolitan IR: A Response to Burke et al.'s 'Planet Politics,' *Millennium: Journal of International Studies*, 46(2), pp. 190–208.

[11] Randazzo Richter, 'The Politics of the Anthropocene,' p. 296. One should also note the use of Indigenous knowledge for bolstering hegemonic governance regimes, which, at times, result in the occlusion of the experiences and struggles of indigenous peoples. See D. Chandler and J. Reid (2019) Becoming Indigenous: Governing Imaginaries in the Anthropocene, Rowman and Littlefield and P. Anthias and K. Asher (2024) 'Indigenous Natures and the Anthropocene: Racial Capitalism, Violent Materialities, and the Colonial Politics of Representation,' *Antipode: A Radical Journal of Geography*, online.

blighting current multi-species relations is indeed the 'commodification' of non-human sentient beings, then this implies that some kind of systemic transformation would be necessary. The chapter touches on the production of 'cheap food' through factory farming and how this constitutes such an intrinsic and devastating dimension of existing multi-species relations; it is difficult to envisage how any substantive change in these relations could take place without first some kind of human-induced systemic change of the global capitalist economy. We thus arrive at the vexed question over whether the ontopolitical assumptions of the author's 'more-than-human' framing of the 'Anthropocene' allows for this. I believe the answer to be in the affirmative, and that this is in fact implied in the chapter. As the author points out, non-human species cannot achieve recognition or rights on their own, and thus we can deduce that this burden logically falls back upon human beings. What remains to be seen is how the systemic conditions of global capitalism might be transformed in ways which lead to the reconfiguration of multi-species relations.

Similar ontopolitical questions arise regarding the transformative agential capacities of human beings in Molinero-Gerbeau and Pérez's chapter. If classical Marxist dialectics designates the proletariat as the agential force capable of bringing about revolutionary change, the Monist dialectics informing the authors' World-ecology framework appears to curtail this since we would seemingly succumb to the 'Anthropocentrism' or 'Cartesian dualism' which advocates are so keen to avoid. Yet one could argue that it is not only advantageous to maintain some distinctions between 'society' and 'culture,' but also unavoidable. Some authors have even pointed out how key proponent Jason Moore's own work shows signs of 'dualist' thinking.[12] It is also difficult to envisage any forms of political praxis without referring to the unique agential capacities of human beings. Indeed, Molinero-Gerbeau and Perez hint at such a thing as their reformulation of the category 'climate migrant' for 'CoPe' (core/periphery mobilities) suggests. Of course, international migration regimes of governance are not solely enacted at the level of discourse but also in material practices, and thus resistance cannot remain at the level of ideological critique but must also entail some kind of performative human agency. What seems to be required is a new research agenda around the Capitalocene 'which includes social struggles, creative practices, and possibilities for different futures; and an openness of the future that takes the notion of politics and alternatives seriously.'[13]

In San Román's chapter, human subjectivity and agential capacities in the 'Techno-Westernocene' are not addressed substantively. Yet, the author's dissatisfaction with conceptions of the 'Anthropocene' and the 'Capitalocene' for failing to escape 'Cartesian dualism' signposts an even more stringent commitment to Monism

[12] See S. Nelson (2016) 'Book review essay- Sara Nelson on Jason Moore's "Capitalism in the Web of Life: Ecology and the Accumulation of Capital,' *Antipode Online*, 9th March, 2016, and A. Hornborg (2020) 'The World-System and the Earth System: Struggles with the Society/Nature Binary in World-Systems Analysis and Ecological Marxism,' *Journal of World-Systems Research*, 26(2), pp. 194–195.

[13] K. Jansen and J. Jongerden (2021) 'The Capitalocene Response to the Anthropocene,' in H. Akram-Lodhi, K. Dietz, B. Engels, and B. McKay (eds) *The Edward Elgar Handbook of Critical Agrarian Studies*, Cheltenham: Edward Elgar, p. 644.

than we find in the other chapters of the volume. The 'Techno-Westernocene' is populated and constituted by what is articulated as the totalizing, all-encompassing and seemingly deterministic force of 'technology.' While one can, and indeed should, recognize the agency of non-human materiality (such as through advanced technologies related to artificial intelligence, the Internet of Things, or new codes and malware[14]), it does not follow that these artifacts have become completely detached and autonomous from the grasp of human beings. The Monism which grounds San Román's framing of the 'Techno-Westernocene' would appear to leave little room for the contestation of political subjectivities or transformative human agency. Yet, as my review of the previous chapters has aimed to show, the commitment to Monist relational ontologies need not lead to the complete dissolution of the human subject and depoliticization. Assuming that San Roman's main thesis holds (we are living in a 'Techno-westernocene'), the unique capacities that human beings possess for critique, contestation, or political praxis can never be entirely absent. Let us return to the conception of 'technology' advanced in the chapter. I have previously suggested that this hints at technological determinism, but, in reality, this is not possible since 'technology' always transcends the material; there must always be a social dimension.[15] San Román's conception of 'technology' is in fact a deeply social one, being derived from a specifically Western 'worldview' and knowledge, meaning that this system has a history and a political one at that.[16] Thus, we can deduce that the conditions of possibility for political contestation and human transformative agency already reside from within the 'Techno-Westernocene.'

Robson's chapter addresses most explicitly the question of political subjectivity and transformative human agency. Through the Deleuzoguattarian 'assemblage' approach deployed the author makes distinctions between different kinds of socio-political assemblages. This, in turn, leads to the advocacy of social movement 'war-machines' as the most effective formation for the contestation and transformation of capitalist states implicated in climate change. On the one hand, Robson highlights the potential of these formations for struggle at the level of discourse deemed necessary to challenge the constitution of passive and depoliticized subjectivities sought by hegemonic forms of 'green' or 'climate' governmentality. This seems necessary to carve out a political space of freedom within which human subjects can pursue an alternative political project. On the other, he stresses their capacities for engaging in performative political action (understood as an indispensable part of political struggle

[14] See, for instance, M. Carr and F. Lesniewska (2020) 'Internet of Things, Cybersecurity and Governing Wicked Problems: Learning from Climate Change Governance,' *International Relations*, 34(3), pp. 391–412 and N. S. Fouad (2022) 'The Non-anthropocentric Informational Agents: Codes, Software, and the Logic of Emergence in Cybersecurity,' *Review of International Studies*, 48(4), pp. 766–785.

[15] See, for instance, S. Wyatt (2008) 'Technological Determinism Is Dead; Long Live Technological Determinism,' in E.J. Hackett et al. (Eds.) *The Handbook of Science and Technology Studies*, Cambridge: the MIT Press, pp. 165–180.

[16] C. Otter (2022) 'Socializing the Technosphere,' *Technology and Culture*, 63(4), pp. 953–978.

in these times of ecological crisis).[17] Human agency is crucial for both of these, it's just that the ontological assumptions grounding the author's assemblage approach posit that this must be immanent and situated within particular assemblages. For all that, one must also recognize that the construction (and efficacy of such assemblages) is undoubtedly more difficult to achieve in practice than in theory. States and the machines of global capitalism possess vast resources to suppress dissent and will inevitably seek to re-territorialize these recalcitrant movements to blunt their transformative capacities. Moreover, these social movement formations would likely need to incorporate what Eugene Holland calls 'Anthropocene citizens' who are' operating within established institutions, both governmental and non-governmental.'[18]

This (albeit limited) survey of the ontopolitics of different framings of the 'Anthropocene' reveals a nuanced, varied, and optimistic outlook vis-à-vis the scope for compelling articulations of dissenting human subjectivities and transformative human agency. As noted in the Introduction to this volume, some authors have quite rightly expressed concerns in this regard.[19] Chief among them concerns the decentering of the human subject, which can lead to the erosion of its unique agential capacities; if human agency becomes compromised to the extent that the possibilities for challenging hegemonic power structures and pursuing transformative political proposals vanish with it then solutions to the climate crisis would surely remain out of reach.[20] However, in this conclusions section I have tried to show how the authors hint at diverging ontological assumptions regarding Monism or relationality, and how this has important political implications for human subjectivity and transformative agency. It is thus not that relational approaches to the 'Anthropocene' (or alternative conceptions) lead inevitably to the dissolving of the human subject and depoliticization in the face of deteriorating ecological context, but rather the ontological assumptions we make concerning the nature of those relations and the place of the human being within them which ultimately matters. Some authors working within IR and Political Geography have already been exploring how, from the ashes of Modernist frameworks, human beings can pursue alternative worlds.[21] These conclusions also concur with Erika Cudworth and Stephen Hobden's insistence that we keep in mind the heterogeneity of posthuman approaches; while there may well

[17] E. Swyngedouw and H. Ernstson (2018) 'Interrupting the Anthropo-obScene: Immuno-Biopolitics and Depoliticizing Ontologies in the Anthropocene,' *Theory, Culture & Society*, 35(6), pp. 3–30; W. E. Connolly (2015) 'Confronting the Anthropocene and Contesting Neoliberalism: An Interview with William E. Connolly,' *New Political Science*, 37(2), p. 267.

[18] E. W. Holland (2016) 'On/Beyond the Anthropocene,' *Deleuze Studies*, 10(4), pp. 564–573.

[19] D. Chandler (2018) *Ontopolitics in the Anthropocene: An Introduction to Mapping, Sensing and Hacking*, Routledge.

[20] D. Chandler (2013) 'The world of attachment? The post-humanist challenge to freedom and necessity,' *Millennium: Journal of International Studies*, 41(3), pp. 516–534.

[21] B. Evans and J. Reid (2014) Resilient Life: The Art of Living Dangerously. Polity; K. Grove (2018) Resilience, Oxon: Routledge.

be valid grounds to denounce the impoverished accounts of human subjectivity and agency in some strands of posthumanism this need not be the case for all of them.[22]

Ontopolitics and Onto-Ethics of the 'Anthropocene'

This Conclusions section has been principally concerned with the ontopolitical commitments inherent in different framings of the 'Anthropocene,' but there is another related dimension I would like to touch on here. Each of the chapters in this volume harbors a certain ethical orientation or implications, which should not be considered a mere appendage or afterthought but rather a fundamental dimension of the ontopolitics of the 'Anthropocene.' That is, just as the authors' ontological assumptions shape the ways in which they engage with politics, so too are they indissociable from the ethical possibilities which open out before them. To be sure, this ethical orientation is more explicit in some chapters than in others. Yet what we seem to find is a convergence between ontopolitics and onto-ethics, suggesting that any consideration of the former must also take note of the latter. I shall now carry out a selective survey of the contributions of this volume, seeking once again to bring them into contact with existing work on ethics in the 'Anthropocene.' We will draw this conclusion chapter to a close by offering some final thoughts on the importance of maintaining a focus on both the ontopolitics and onto-ethics of the 'Anthropocene.'

In the chapters by Pereira et al. and Kuetting the authors' relational ontology underpinning their conception of the 'Anthropocene' leads them to consider the move from a more anthropocentric ethics to one which is 'beyond the human.' This is perhaps most pronounced in Kuetting's exploration of an ethics of 'multi-species justice.' As the author points out, the fact that non-human sentient beings are devoid of agency (at least in a deliberative sense) does not necessarily mean that they should be denied rights, recognition, or indeed justice. The ethical responsibility to pursue multi-species justice in the 'Anthropocene' therefore falls back on human beings themselves. In addition, just as the ontological assumptions infused in our framings of the 'Anthropocene' can affect the political attribution of rights and recognition for non-human sentient beings, the same appears to ring true for the ethical question of multi-species justice. Rethinking multi-species relations via a relational approach to the ethics of multi-species justice takes us away from the human-non-human binary and the individual (human) subject of rights to embrace the relations of multi-species 'encounters' and the reconceived ethical responsibilities that this entails.[23]

In some of the other chapters, the ethical commitments would appear to be less evident, but upon closer inspection they are not absent. Such is the case with

---

[22] E. Cudworth and S. Hobden (2015) 'Liberation for Straw Dogs? Old Materialism, New Materialism, and the Challenge of an Emancipatory Posthumanism,' *Globalizations*, 12(1), pp. 134–148.

[23] P. Tschakert (2020) 'More-Than-Human Solidarity and Multispecies Justice in the Climate Crisis,' *Environmental Politics*, 2 (31), pp. 277–296.

Molinero-Gerbeau and Pérez's deployment of a World-ecology framework and the Monist relational ontology on which it is grounded. As Jason Moore puts it, World-ecology is not devoid of ethics since it 'privileges an engaged pluralism and an *ethics of synthesis* committed to building the international solidarities necessary to effectively resist-and thence to socialize-the International of Capital.'[24] This 'ethics of synthesis' is perhaps best understood as an ethic of critique and re-composition, whereby through a certain dialectics the pernicious Cartesian dualisms employed by capitalism can be dissolved and reformed with the totality of the whole recovered. In Molinero-Gerbeau and Pérez's chapter, we can see how the authors are clearly not only interested in critiquing and dismantling the culture-nature and legitimate-illegitimate dualisms in which they maintain structure international migration regimes in the interests of global capitalism. Through their pluralistic approach to mobility, the ethical imperative is also to expose and address the injustices which these migration regimes hold in place by reformulating the pernicious categories into something different—'CoPE' (Core-Periphery) mobilities.

The ethical undercurrent of Robson's chapter is similarly latent but not absent entirely. To grasp this, it is necessary to return to the relational ontology underpinning the author's Deleuzoguattarian assemblage approach. The work of these authors is of course not known for its normative disposition, yet, as some have noted elsewhere, one can still glean a kind of immanent ethics which privileges the Nietzschean concern with increasing one's power, or more precisely one's capacities to act.[25] If it is indeed the case, as Robson argues, that 'green' and 'climate' governmentalities restrict human capacities vis-à-vis the contestation of capitalist states and the mitigation of climate change, then the production of 'lines of flight' by social movement war-machines to increase the capacities of human beings could be viewed in ethical terms. In so far as these operations can lead to the liberation from restrictive regimes of governance and the transformation of destructive political-economic structures then one can surely discern a clear ethical orientation.

The chapters in this volume thus exhibit how the ontological assumptions grounding different framings of the 'Anthropocene' not only foreground the possibilities for varying political proposals but also have a bearing on the kind of ethics that we can pursue. Within the broader literature on climate change, we can find diverse configurations of ontopolitics and onto-ethics. For instance, authors using anthropocentric liberal frameworks have been advancing rights-based approaches to advocate for 'distributive justice.'[26] From what could be deemed a similarly anthropocentric eco-socialist perspective, it is argued that any pursuit of justice must take

---

[24] J. Moore (2022) 'Anthropocene, Capitalocene & the Flight from World History: Dialectical Universalism & the Geographies of Class Power in the Capitalist World-Ecology, 1492–2022, *Nordia Geographical Publications*, 51(2), p. 138, my emphasis added.

[25] D. W. Smith (1998) 'The Place of Ethics in Deleuze's Philosophy: Three Questions of Immanence,' in E. Kaufman and K. J. Heller (eds) *Deleuze and Guattari: New Mappings in Politics, Philosophy, and Culture*, Minneapolis: University of Minnesota Press.

[26] S. Caney (2018) 'Distributive Justice and Climate Change,' in S. Olsaretti (ed) *Oxford Handbook of Distributive Justice*, Oxford University Press.

into consideration the injustices of global capitalism.[27] However, the contributions made in this volume support existing research endorsing the ethical possibilities inherent in relational approaches to the 'Anthropocene.' As noted elsewhere, relational ontologies can potentially augment ethical commitments since they invariably highlight the interconnections between human beings, other living beings, and our natural world.[28]

I would like to bring this volume to a close by offering some final comments on the ethical dimension of the ontopolitics of the 'Anthropocene.' In a transdisciplinary literature, authors have been turning to the ethics of Emmanuel Levinas or Jacques Derrida to articulate our responsibility or 'hospitality' for the Other in these times of profound ecological change.[29] Such ethical frameworks seem like a particularly good fit with the kind of relational ontologies we have been exploring in this volume, and indeed some of the authors, albeit implicitly, seem to venture in this direction: e.g., whether we speak about our responsibility to each other as human beings, to migrant Others, to non-human sentient Others, or indeed to the natural world as the ultimate Other. Nonetheless, once more I would like to underline the importance of not jettisoning the human subject or human agency, or disavowing its unique ontological status in favor of drawing equivalences with other forms of materiality. Accepting that responsibility to the Other depends on an irreplaceable 'Self' necessitates maintaining a certain tension between the 'Self' and 'Other.'[30] Dan Bulley has made a similar point when cautioning against dissolving our 'ontopological' situatedness in favor of a kind of pure multiculturalism free from subjective ties to territory, patria, or space.[31] Such subjective notions of the Self, of our own ontopological place in the World, are the very conditions of possibility for 'hospitality' and thus cannot be effaced. Thus, relational ontological assumptions grounding framings of the 'Anthropocene' may well foment our ethical commitments to a range of Others, but for that a situated conception of human subjectivity and agency must still be articulated.

To conclude, let us re-route this discussion back to the politics of the 'Anthropocene,' since, as Erik Swyngedouw and Henrik Erntson have argued forcefully, ethics and philosophy will never be enough to save us.[32] Professing responsibility or 'hospitality' to the Other is indeed inspiring, but without engaging in performative political struggle and contestation this will likely be of little significance.[33] Future work on relational approaches to the 'Anthropocene' and IR could perhaps

[27] D. Faber, B. Levy, and C. Schlegel (2021) 'Not all People are Polluted Equally in Capitalist Society: An Eco-Socialist Commentary on Liberal Environmental Justice,' *Capitalism Nature Socialism*, 32(4), pp. 1–16.

[28] E. Cudworth and S. Hobden (2018) *The Emancipatory Project of Posthumanism*, Routledge.

[29] N. Clark (2011) *Inhuman Nature: Sociable Life on a Dynamic Planet*, Sage.

[30] V. Paipais (2011) 'Self and Other in Critical International Theory: Assimilation, Incommensurability and the Paradox of Critique,' *Review of International Studies*, 37, pp.121–140.

[31] D. Bulley (2006) 'Negotiating Ethics: Campbell, Ontopology and Hospitality,' *Review of International Studies*, 32, pp. 645–663.

[32] Swyngedouw and Erntson, 'Interrupting the Anthropo-obScene'.

[33] Ibid.

explore the connections between ontopolitics and onto-ethics: e.g., how the formulation of ontopolitical proposals implies a sense of responsibility for the Other, while, conversely, how the articulation of responsibility for the Other entails the need for political struggle via the constitution of dissident subjectivities and the elaboration of situated forms of human agency.

The manufacturer's authorised representative in the EU is Springer Nature Customer Service Centre GmbH, Europaplatz 3, 69115 Heidelberg, Germany. If you have any concerns regarding our products, please contact ProductSafety@springernature.com

Printed and bound by CPI Group (UK) Ltd, Croydon, CR0 4YY
01/10/2026
02225876-0015